Palgrave Perspectives on Process Philosophy

Series Editors
Wahida Khandker
Department of History, Politics and Philosophy
Manchester Metropolitan University
Manchester, UK

Tim Flanagan
Philosophy
Murdoch University
WA, Australia

The aim of this series is to provide monographs, edited collections and Palgrave Pivots from both established and early career scholars in Process Philosophy, with particular reference (but by no means exclusively) to the writings of Alfred North Whitehead, Henri Bergson, William James, and Charles Hartshorne. The series aims to promote new writing in this area that is innovative, rigorous, and yet accessible to readers both within philosophy and beyond disciplinary boundaries.

The expected focus of the books falls roughly into three areas, though it is likely that these areas will overlap:

- New readings of the major early twentieth-century process philosophers, Whitehead, Bergson, James, Hartshorne, et al, including studies of their influence on, or relevance to, contemporary philosophy
- The roots of process philosophy in ancient and early modern philosophy, as well as hitherto unexplored affinities with thinkers in these eras
- Future orientations: including cross-disciplinary explorations of the implications of process thought. We are particularly interested in studies of intersections with the arts, architecture, environmental humanities, and the sciences, but projects proposing other disciplinary intersections will be welcomed.

More information about this series at
http://www.palgrave.com/gp/series/16106

Wahida Khandker

Process Metaphysics and Mutative Life

Sketches of Lived Time

palgrave
macmillan

Wahida Khandker
Department of History, Politics and Philosophy
Manchester Metropolitan University
Manchester, UK

ISSN 2524-4728 ISSN 2524-4736 (electronic)
Palgrave Perspectives on Process Philosophy
ISBN 978-3-030-43047-4 ISBN 978-3-030-43048-1 (eBook)
https://doi.org/10.1007/978-3-030-43048-1

Cover illustration: Hongbo Liu/Moment/Getty

This Palgrave Macmillan imprint is published by the registered company Springer Nature Switzerland AG
The registered company address is: Gewerbestrasse 11, 6330 Cham, Switzerland

Series Editors' Preface

If the European philosophical tradition can be characterised as 'a series of footnotes to Plato' (Whitehead 1978, p. 39) an attendant claim might well be that the work of Whitehead has, in its own right, come to constitute an *opus citatum* for subsequent thought. To say as much is not simply to see Whitehead's project as if it were some sort of Whig historical emendation (a modern addendum to ancient corrigenda); rather, to read his thought is to encounter, *passim*, an ongoing engagement with classical themes and problems from inherited antiquity. This much at least Whitehead, whose 'organic philosophy only repeats Plato' (Whitehead 1978, p. 94), explains when he insists that while 'we note the many things said by Plato in the *Timaeus* which are now foolishness, we must also give him credit for that aspect of his teaching in which he was two thousand years ahead of his time' (Whitehead 1978, p. 94).

Long before Whitehead, an earlier reference in the very same 'series of footnotes,' though one to be found in the same place, *ibidem*, *loco citato*, is provided by Plotinus. In his disquisition 'On the Three Primary Hypostases' (that has in turn come to characterise the reception of his project[1]) Plotinus declared that his own statements are mere 'exegeses' [λόγους ἐχηγητάς] of what can be found in Plato, albeit 'not explicitly [μὴ ἀναπεπταμένως]' (Plotinus 1989, pp. 40–41). For, as developed elsewhere in the important treatise 'On Eternity and Time,' 'although we should consider that some of the ancient and blessed philosophers have said what is right, we ought to examine which ones came closest and *how we, too, can attain comprehension* [καὶ πῶς ἂν καὶ ἡμῖν σύνεσις περὶ τούτων γένοιτο] regarding these matters' (emphasis added) (Plotinus 1989, pp. 296–297).

In this way, for Plotinus as much as for Whitehead, to engage the works of Plato is an undertaking that requires these texts be as much 'read' as *actually* (that is, as if for the first time) 'thought.' The reason for this is that the textual or paleographic question of exactly '*what* is to be thought' is one that serves as a transcendental rubric for the ontological question of quite 'what thought is *to be*.'[2] As such, the 'conception of an actual entity in the fluent world' (Whitehead 1978, p. 82) sought by Whitehead might be seen as a glimpse of that enigmatic 'image of Eternity' from the *Timaeus*, that very 'image' which Plotinus sought to experience through 'continuous application' of our concepts.

To (actually) conceive of this fluency on its own terms rather than according to static categories is, as Whitehead acknowledged, precisely what Bergson, one time Collège de France Chair of Greek and Roman Philosophy, undertook in his own reading of the history of philosophy. For upon the Bergsonian account of *durée*, it follows that the comprehension of a text proceeds not according to an external order but rather to a sense of understanding that only emerges *in* the act of reading. And even if on this Bergson departed in significant ways from certain aspects of Platonism, what this shows is that the source of any 'intellect' is more an ongoing process than an atemporal doctrine.

---

[1] For a recent discussion of the organising role of hypostases in scholarship on Plotinus, particularly on whether the threefold set of principles is to be regarded as having real (ontological) weight or merely furnishing an interpretive (systematising) rubric, see Catana (2013).

[2] See the work of Barbara Cassin, after Novalis, on logology.

'Sensation,' Bergson writes, in his essay on the nature of dreams, 'is warm, coloured, vibrant and almost living, but vague; memory is clear and distinct, but without substance and lifeless' (Bergson 1975, p. 118). Bergson finds here an analogy with a passage from Plotinus' *Enneads* on the work of Nature, whose incomplete 'sketches' of living bodies are only completed in the aspirations of souls towards embodiment. In our dreaming state, our memories seek matter to realize them, detached from sensations and actions that might define them. Without sensations, these 'phantom memories' play out their lives in dream. In our waking state, knowledge itself is a continuous effort to bring memories to bear on a world sketched out for us by our senses and actions. In the light of this Plotinian metaphor for the work of the intellect Bergson also speaks of the role of the 'dynamic scheme.' If images, in the sense that Bergson intends, are the coincidence of the past and present, or the work of memory intervening in perception in the process of recollection, then the dynamic scheme is an image-in-the-making, in the process of invention. In the interconnected acts of reading, interpretation, and writing we realise what commences only as an imperfectly envisaged idea, whilst the work only attains its distinctiveness and novelty (as opposed to the reconstitution of existing pieces in the manner of a mosaic) in *the time it takes* to bring it to completion.

Why have we commenced with this general statement about the nature of philosophizing as an unfolding, creative activity? It is, quite simply, in the spirit of this interest in the possibilities of philosophical writing itself that we proposed this series, a proposal whose organizing principle is one oriented by philosophical processes rather than by philosophical doctrines. It is our hope that each book in the series will serve as a provocation and an invitation to such new avenues of research in process philosophy. We anticipate that manuscripts submitted for the series will seek innovative ways to bridge gaps between the varied traditions of process thought and historical, contemporary, and cross-disciplinary problems. Proposals for monographs, 'Pivots,' and edited collections might examine some aspect of the following areas of research and beyond.

1. The recent resurgence of interest amongst philosophers in metaphysical accounts of dynamism, drawing on the history of philosophy:
Beyond the perennial influence of Pre-Socratic cosmology, the more recent traditions of eighteenth-century German Romanticism, nineteenth-century French Spiritualism, and early twentieth-century Process Metaphysics have each in their own way underlined the sense in which a great many of the ancients' conceptual schemes continue to merit close attention, even if many of their empirical claims have been outmoded. So too, while the investigation into 'dispositions' in contemporary metaphysics serves to propose an account of the physical world, in the end the naturalism that this entails is often set out in terms of an Aristotelian understanding of categories and predication.[3] On the other hand, an engagement with Platonic or Neo-Platonic themes such as participation, emanation, and generation is one that operates by way of a consideration of nature that is *itself* ontological rather than a modality of ontology. Such thought provides for a consideration of nature that is oriented by actual events and living things, rather than by definitional accounts of their formal essence.[4]

2. Perspectives on the growing cross-disciplinary awareness and uptake of process thought as a means of thinking a new problems in art practice, the history of architecture and design:
There is, for example, emerging interest in kinaesthetics in both art criticism and the theoretical reflections of artists themselves concerning their own practices. This invites a renewed focus on the work of a number of process philosophers for their unique and influential accounts of embodied experience and fragmented subjectivity. Process philosophy is notable in this respect, first, historically for its direct influence upon Continental philosophers typically renowned for their studies of embodiment such as Maurice Merleau-Ponty, Gilles Deleuze, and more recently, the broad area of 'New Materialism.' It is significant, secondly, for Bergson's still

[3] See Groff and Greco (2013) and Marmadoro (2010).

[4] See, for example, Wagner (2001), Hadot (2006), Wilberding and Horn (2012).

distinctive dissolution of the mind-body dichotomy as a means of re-thinking the very nature of matter no longer in terms of space but rather in terms of time, and Whitehead's fragmentation of subjectivity into multiple levels of process analysable into microscopic acts of perception that he names 'prehensions.' This series encourages research into the historical setting and implications of process thought for an understanding of artistic practices (drawing, painting, sculpture, etc.) as explorations of the processive nature of reality.

3. The histories of the physical and biological sciences and the growing dynamism of scientific theories in the twentieth century:
Process Philosophy is becoming increasingly relevant as a frame of reference for our understanding of the mind, biological systems, and ecosystems. Future monographs and collections might examine connections between process philosophy and biological and physical concepts of change and development, both in the history of science and in contemporary scientific research. If the philosophy of biology, as Ernst Mayr observed, has been obscured by an increasingly positivist impetus to reduce scientific knowledge into principles most resembling physico-chemical structures, then the proliferation of process perspectives might help to enrich and strengthen the philosophy of biology. Furthermore, in the light of recent shifts in the mainstream political awareness of the problem of anthropogenic climate change, it seems timely to consider the ways in which process philosophy can help us to articulate and understand this problem and its resulting environmental and social impacts. How might philosophical, cosmological, and cosmogonical concepts (of change, displacement, processes of othering, memory, and renewal) intersect with geographical, political, and legal discourses on the displacement of both human and more-than-human populations?

WA, Australia — Tim Flanagan
Manchester, UK — Wahida Khandker

References

Bergson, Henri. 1975. *Mind-Energy: Lectures and Essays*, trans. H. Wildon Carr. Westport, CT: Greenwood Press.

Catana, L. 2013. Changing Interpretations of Plotinus: The 18th-Century Introduction of the Concept of a 'System of Philosophy.' *The International Journal of the Platonic Tradition* 7: 50–98.

Groff, Ruth, and John Greco, (eds.). 2013. *Powers and Capacities in Philosophy: The New Aristotelianism.* London: Routledge.

Hadot, Pierre. 2006. *The Veil of Isis: An Essay on the History of the Idea of Nature*, trans. Michael Chase. Cambridge, MA: Harvard University Press.

Marmadoro, Anna, (ed.). 2010. *The Metaphysics of Powers: Their Grounding and Their Manifestations.* London: Routledge.

Plotinus. 1989. *Plotinus: Enneads*, trans. A.H. Armstrong., Cambridge, MA: Loeb Classical Library.

Wagner, Michael F. 2001. *Neoplatonism and Nature: Studies in Plotinus' Enneads.* New York: SUNY Press.

Whitehead, Alfred North. 1978. *Process and Reality: An Essay in Cosmology.* New York: The Free Press.

Wilberding, James, and Christoph Horn, (eds.). 2012. *Neoplatonism and the Philosophy of Nature.* Oxford: Oxford University Press.

Acknowledgments

I am grateful to the History Research Centre at Manchester Metropolitan University for supporting a short period of research leave that enabled me to work on this project. Thanks also to Brendan George at Palgrave Macmillan. Special thanks to Liz Tyson, Nicola O'Brien, Eva Giraud, Louise Pallister, and John Ó Maoilearca, and I am most indebted to Tim Flanagan for a collaboration, spanning much of the time spent researching and writing this book, that has helped to bring this work to fruition.

Contents

1
Incipience

It is June 2016, and a report by the state government of Queensland, Australia, declares that a species of rat, *Melomys rubicola*, or the Bramble Cay melomys, is in all likelihood now extinct (Watson 2016). Its decline and eventual extinction had been precipitated by repeated inundations of its small island habitat in Torres Strait. The increased frequency of the inundations is in turn thought to be due to climate change, exacerbated by human activity, making this mammalian extinction the first to be attributable to this particular cause. For this reason, this most 'uncharismatic' of animals has, *post mortem*, achieved a level of prominence in the public eye that it consistently failed to gain over the course of its acquaintance with *Homo sapiens*. Had this small rat, a relative of that most familiar of species *Rattus rattus* that has for the longest time been labelled a pest, lived somewhere more directly accessible to humans, and looked more appealing, the signs of its rapid decline over the previous decades might have excited greater concern amongst those agencies responsible for the preservation and promotion of biodiversity and amongst the general populace (Fulton 2017).

Around three years later, in April 2019, *The Astrophysical Journal Letters* publishes results from the Event Horizon Telescope Collaboration

W. Khandker, *Process Metaphysics and Mutative Life*,
Palgrave Perspectives on Process Philosophy,
https://doi.org/10.1007/978-3-030-43048-1_1

to view the 'shadow' of a supermassive black hole (one that is millions of times the size of our Sun) theorized to occupy the centre of a galaxy named 'M87'. The unveiling of the images is broadcast worldwide, and the collaboration itself is international in scope, gleaning data from an array of telescopes, and research teams, scattered across the globe. The images, once processed, indicate the presence of a black hole spinning in a clockwise direction, roughly as envisaged in line with the predictions of General Relativity.[1] What looks like a 'corona' of light around a black circle is in fact an accretion ring, and the ring's skewed orientation indicates the phenomenon's direction of spin. This movement biases the accumulation of debris towards its 'south' due to a combination of gravitational lensing and relativistic beaming (Collaborat, Event Horizon Telescope 2019, p. 6). The 'hole' in the centre of this image is merely the *shadow* of a black hole, the elusive object itself, the singularity hidden behind the event horizon, possessing such mass that even light cannot escape its gravitational pull.

In short, in the time it has taken to write this book, we have been confronted with the first mammalian extinction caused by anthropogenic climate change. Behind this event, and advancing even more imperceptibly, our deliberate and incidental effects on our environment are accelerating natural extinctions to anything from 100 to 1000 times their normal 'background rate' (Woinarski et al. 2017, p. 14). Yet we have also witnessed the first images of a region of space surrounding a black hole, an awe-inspiring and thus equally humbling event. Both examples serve

[1]The team describes the appearance of the black hole, compared to simulations modelled in conformity with key principles of General Relativity, such as the effect on light in the presence of immensely strong gravity, known as 'gravitational lensing.' Illustrating with two variants of this, they explain: 'The simulations of Luminet showed that for a black hole embedded in a geometrically thin, optically thick accretion disk, the photon capture radius would appear to a distant observer as a thin emission ring inside a lensed image of the accretion disk. For accreting black holes embedded in a geometrically thick, optically thin emission region, as in LLAGNs, the combination of an event horizon and light bending leads to the appearance of a dark "shadow" together with a bright emission ring that should be detectable through very long baseline interferometery (VLBI) experiments. Its shape can appear as a "crescent" because of fast rotation and relativistic beaming' (Collaborat, Event Horizon Telescope 2019, p. 2). I will return to discussions of the physics, and its philosophical implications, underpinning knowledge of black holes later in the chapter 'Reminiscence' which was written before the results of the Event Horizon collaboration were published.

as a reminder of our collective power, both to care and to harm, and ultimately of the fragility of all living beings, human or otherwise, compared to the vastness of the universe that encompasses us. The study of endangered species on Earth and black holes in space both seek to make perceptible the most elusive of scientific objects of analysis. The problem of imperceptible processes is indeed one concerning the nature of scientific discovery and the role of metaphysics, but it is fundamentally also one that informs the manner in which we live on a planet under stresses for which we, as a species, are responsible. Much of this book is devoted to the details of philosophical and biological research, but always against the background of our increased awareness of environmental crisis and changing attitudes towards our fellow species on Earth, provoked by new perspectives, and ever-renewing techniques to generate such perspectives, on living processes and their complex interactions.

The structure of this book falls into three parts together providing a survey of key process philosophical approaches that, in conversation with selected concepts across the biological and physical sciences, help us to think about living processes, or 'lived time,' at different scales of functioning. The first part is written from an opening perspective on the question of the differing *scales* of analysis provided by Alfred North Whitehead. In particular, his interest in questions arising from the quantum mechanical reconciliation with classical mechanics informs the first two chapters that address problematic categorizations of life as variously 'despotic,' 'invasive,' or as primitive (in the radically more-than-human case of micro-organisms), whose potential recategorization relies on our willingness to acknowledge changes in value depending on the scale at which we view them. The second part of the book concerns *methodologies,* in the light of works by Henri Bergson, whose intertwining concerns with epistemology and ontology in his theories of mind and life serve as a model for a process philosophy of biology. The focus there will be on techniques used across philosophy and the sciences to visualize processes that are otherwise unavailable to us due to the limitations of our perceptual faculties, no matter how sophisticated the tools for analysis, from microscopes to telescopes, have become. This book concludes with a consideration of the *relations between parts and wholes* in process, panpsychist, and ecological terms. It revisits the question of ecological balance and the place of human activities in relation to it, with reference to works of Charles Hartshorne and William James. Before I summarize the aims of each chapter in more detail, I want to provide an overview of some of the historical and philosophical engagements with the sciences that have informed this book.

1 Process and Biology

Bergson opposed images and concepts, the tools of analysis, to an ideal of 'intuition,' the latter being both a faculty (in response to Kant's identification of the faculties that shaped knowledge) and the practice of metaphysics itself. Take as many views of an object from without as one wishes, and one will never attain a complete knowledge of that object. Whatever it is, it is so internally, in such a way that symbols and points of view on it are necessarily locked outside. 'Description, history, and analysis' are merely ways for us to achieve relative knowledge (Bergson 1992, p. 160). Absolute knowledge, on the other hand, can be gained through an intuition, or 'the sympathy by which one is transported into the interior of an object in order to coincide with what there is unique and consequently inexpressible in it' (Bergson 1992, p. 161). A distinction between science and metaphysics presents itself here. For what Bergson calls positive science, the analysis of things in terms of symbols is its *modus operandi.* 'Even the most concrete of the sciences of nature, the sciences of life,' he continues, cannot help but reduce the complex living forms it encounters, the complex interactions and processes internal to each organism's body, and the ecosystems within which they become entangled, to systems of simpler terms, that constitute their 'visual symbol' (Bergson 1992, p. 162). Yet the remedy to this reliance upon symbolical representation, a metaphysics that is 'the science which claims to dispense with symbols' (Bergson 1992, p. 162), is a convergence of efforts from many different areas of study in a kind of kaleidoscopic multiplication of viewpoints on one's chosen object. The key to this multiplication of images is their difference from one another. Call on images that are too alike, and one succumbs to the will to reduce everything to a common symbol. Choose reference points that are sufficiently distinct, and the effect of dissimulation is to intensify the sense of one's own intellectual effort being exercised in the course of the enquiry. It is less a matter of defining and reducing the object to a familiar concept, than it is of directing one's efforts of enquiry in a more fruitful direction (Bergson 1992, p. 166). This summary of Bergson's account of metaphysical enquiry is not intended as obfuscation, but rather to accentuate the importance of *method* as the goal of metaphysics, since the definition of one's object of study can only be achieved through agreement upon a set of parameters, concepts, and symbols.

The centrality of method starts with an observation that 'duration' is most immediately available to our minds, in the deceptively simple experience we have of our own conscious activity. Once we try to analyse that consciousness into its constituent parts, we both alienate ourselves from it and gain some insight into some of its activity. That is, the tendency to break things into immobile units, as we tend to do with conscious *states*, tells us something about the function of the human intellect. It is directed towards actions that might be useful for this or that purpose. What is most clear and distinct to our minds is simply the set of concepts that we find to be the most useful: 'That is why immobility seems clearer to it than mobility, the halt preceding movement' (Bergson 1992, p. 182). The difficulty arises in the application of such concepts to all aspects of reality that are better understood as 'tendencies.' Again, through inspection of our own conscious processes, we can immediately feel the mind's ceaseless and seamless transition between intensities of concentration and relaxation. However, it is not that the sciences have somehow altogether missed this processual reality, and certain branches, as Bergson identifies, such as mathematics endeavours

> to substitute for the *ready-made* what is in process of *becoming*, to follow the growth of magnitudes, to seize movement no longer from outside and its manifest result, but from within and in its tendency towards change, in short, to adopt the mobile continuity of the pattern of things. (Bergson 1992, p. 190)

The problem with the sciences, as Bergson's diagnosis continues, echoes the problem of the intellect in general. It tends to seek the clarity and distinctness of things once immobilized, and reifies those immobile concepts into an ideal of expression, forgetting that its concepts are the products of the intellect's mode of operation and not of a close contact with the object of study (Bergson 1992, pp. 191–192).

What, then, is the alternative? Bergson's call to create 'fluid concepts' is well-known, thanks in large part to the work of Gilles Deleuze. The potential to follow such a project could be most conspicuously explored in the arts and literature, whilst its less obviously modest potential lies in all efforts of invention across disciplines, including the sciences. Freeing ourselves from the confines of 'inflexible and ready-made' concepts is the *endeavour* of every student of philosophy in their attempts to think through, for themselves, the systems and speculations of the philosophical works presented to them. I emphasize the word *endeavour* both as a caution against hubris and as a description of the real effort at hand, if metaphysics is itself a process, a kind of 'spiritual auscultation' that seeks the closest proximity to the real as possible (Bergson 1992, p. 172). The resulting 'fluid concept' must be specific to its object to such a degree that it is *barely* a concept. It is

> an effort, often painful, immediately to place oneself in the very heart of the subject and to seek as deeply as possible an impulsion which, as soon as found, carries one forward of itself. This impulsion once received sets the mind off on a road where it finds both the information it had gathered and other details as well; it develops, analyzes itself in terms whose enumeration follows on without limit; the farther one goes the more is disclosed about it; never will one manage to say everything; and yet, if one turns around suddenly to seize the impulsion felt, it slips away; for it was not a thing but an urge to movement, and although definitely extensible, it is simplicity itself. (Bergson 1992, p. 199)

To what extent might this metaphysical activity be extended to a philosophy of science? This book asserts an approach to the philosophy of *biology* that is, first and foremost, rooted in a concern with our relationships with other species, not in terms of some separate specialism that we term 'ethics,' as if the good or harms we inflict on other species are 'softer' questions that ought to be debated in toothless ethics committees, whilst biological research ought to be carried out at the precise and disinterested edge of a scalpel. Bergson's philosophy is frequently relegated to a history of vitalism that biologists and philosophers now wholly reject (Mayr 1988, p. 40), but its force lies in its communication of a genuine and joyful inquisitiveness about how life functions, and how we might encounter it.

Why should a concern with the concept of life excite such animated hostility? The historian of science, Georges Canguilhem notes that all 'medical philosophies' after the Aristotelian identification of the 'vegetal state' as the 'minimal expression of the soul's functions,' and that held life to be a unique principle or 'somehow associated with soul' were merely revisiting Aristotle's formulation of the question: 'What is life?' (Canguilhem and Delaporte 2000, p. 67). Canguilhem then sketches out a series of permutations of the concept of life across the eighteenth and nineteenth centuries, from key figures such as Buffon and Linnaeus whose classificatory systems dispensed with the need to identify what 'life' was in favour of a classification of the forms that it took (Canguilhem and Delaporte 2000, p. 68). On the other hand, the physician, Georg Stahl 'insisted that a theory of life was a necessary prerequisite of medical thought and practice' (Canguilhem and Delaporte 2000, p. 68), whilst Xavier Bichat reformulated life as 'the sum of the functions, by which death is resisted' (Bichat 1800, pp. 9–10). Into the nineteenth century, Lamarck 'conceived of life as a continuous, steady accumulation of fluids by solids, initially in the form of cellular tissue, "the matrix of all organization"' (Canguilhem and Delaporte 2000, p. 69), whilst Georges Cuvier saw life and death, no longer as opposing terms, but rather as 'modes of life,' such that 'death is present in life, as both universal armature and ineluctable fate of individual components organized into compatible yet fragile systems' (Canguilhem and Delaporte 2000, p. 71). By the twentieth century, the question 'What is life?' becomes one that even physicists feel able to ask with legitimacy, Canguilhem remarks wryly, citing Erwin Schrödinger's short book bearing that title.

Canguilhem comments on the concern of 'materialist science' with the principle of inertia and thus the configuration and motions of matter abstracted from living organisms themselves, but it is a tendency that continues to be met with resistance. He reminds us that our discussions of animality and its opposite term, the *in*animate, make use of terms whose etymology is rooted in an ancient equivalence drawn between life, soul, and breath ('*anima* = *anemos*') (Canguilhem and Delaporte 2000, p. 74). As Margulis and Sagan observe about the impact of Darwin's theory of evolution by natural selection, animality is sufficiently complex and exceptional for human beings to claim pride in it as their principal feature. Yet we display the opposite tendency 'to shift attention away from the embarrassing fact of our animal ancestry by accentuating those aspects of science that are more akin to physics' (Donald Griffin, cited in Margulis and Sagan 1995, p. 122). The tendency towards the increasingly microscopic did, of course, stem from changes in tools and thus modes of analysis. The continuation of Canguilhem's brief overview of the historical shifts in the epistemology of science emphasizes the effects of the use of the microscope in the latter half of the seventeenth century, one that precipitated a dawning interest in the hitherto unseen complexity of the internal structure of plants and animals:

> The discovery of animalcules, henceforth called Protista, opened up previously unsuspected depths in the empire of the living. Whereas seventeenth-century mechanics was a theory of movements and impulses, that is, a science based on data accessible to sight and touch, microscopic anatomy was concerned with objects beyond the manifest and tangible. Availing oneself of that structural microcosm, that 'other world' within, one could conceive of ever more minute microcosms embedded one within the other. (Canguilhem and Delaporte 2000, p. 79)

He concludes with a consideration of the impact of cybernetic theory on the concept of life, now no longer limited to its concept within the bounds of Bergson's thought that was necessarily limited by the scientific concepts available in the early decades of the twentieth century. Under the aegis of cybernetics, life can be conceived as 'an open dynamical system that seeks to preserve its equilibrium and counteract perturbations.

Such a system is capable of altering its relation to the environment from which it draws its energy in order to maintain the level of some parameter or to perform some activity' (Canguilhem and Delaporte 2000, pp. 86–87). As we will see, the chemist James Lovelock will name the global version of this process *homoeostasis* or, more evocatively, 'Gaia.'[2]

Ernst Mayr's assessment of the prospects of re-invigorating a philosophy of biology (in *Toward a New Philosophy of Biology*, published in 1988) takes place against the backdrop of an increasingly physicalist tradition of the philosophy of science from the 1920s to the 1960s (Mayr 1988, p. 1). Mayr's own response to what Canguilhem called materialist science is to cite the recognition within the sciences of the shift from an understanding of a *universal* application of laws in nature to a *statistical* one. The predominance of stochastic processes at all levels is becoming increasingly manifest 'from subatomic particles to weather systems to ocean currents to galaxies' (Mayr 1988, p. 12). A key obstacle to securing the credibility of biology is, he thinks, a persistence of vitalism, and in contrast to Canguilhem's assessment, it is not a question of the renewed legitimacy of the question, 'What is life?' but the ever-present risk of vitalistic thinking, the remnants of which are found amongst the quantum physicists (Bohr, Schrödinger, Heisenberg) who speculated on the existence of supplementary physical laws that might govern living systems (Mayr 1988, p. 13). A renewed incorporation of biology into the philosophy of science

> will have to abandon its loyalty to a rigid essentialism and determinism in favour of a broad recognition of stochastic processes, a pluralism of causes and effects, the hierarchical organization of much of nature, the emergence of unanticipated properties at higher hierarchical levels, the internal cohesion of complex systems, and many other concepts absent from—or at least neglected by—the classical philosophy of science. (Mayr 1988, p. 21)

[2] I will return to this set of problems in the chapters, 'Symbiosis' and 'Plasticity.' For an initial overview of the themes treated there, see Sects. 2 and 4 of the present chapter below.

In order to become adequate to this array of processes, Mayr proposes a revised conception of teleology that remains in tune with physico-chemical explanations. Thus, any process 'whose goal-directedness is controlled by a program,' that occurs 'only in organisms and man-made machines and constitutes a clear cut difference between the levels of complexity in living and inanimate nature' can be designated as *teleonomic*[3] (Mayr 1988, p. 60).

So far, in this discussion of the idea of a philosophy of biology, I have turned to the work of Bergson and Canguilhem as two of the most significant thinkers of life, both of whom presented challenges to the complete rejection of 'vitalism,' a term that continues to excite hostility and derision, even in the face of attempts to reformulate its meaning. Gary Gutting's survey of the history of what might be called a *Continental* philosophy of science grants both Bergson and Canguilhem their place as leading thinkers within this succession of schools of thought with varying approaches to scientific enquiry. Gutting's initial categorization of three broad approaches to the philosophy of science distinguishes an empiricist/positivist outlook according to which science is the highest form of knowledge, a Kantian/critical stance in which science is the only first-order knowledge and philosophy is tasked with deriving the conditions of possibility of that knowledge, and a metaphysical/ontological view of reality on which science and philosophy are distinct and independent views (Gutting 2005, pp. 1–2). Whatever 'Continental' philosophy of science is, it emerged later than much of the work carried out in line with the approaches listed above. Reflecting a continuing intransigence within both continental and analytic camps, Gutting's diagnosis of the problems of the division between them highlights a particular loss to the enrichment of academic research otherwise promised by intellectual pluralism. Of the state of philosophy in the century preceding the 'break,' Gutting observes that

> even the deepest philosophical divisions (say between a Bergson and a Poincaré or between the early Husserl and the early Carnap) did not

[3] Mayr engages with, and dismisses, Jacques Monod's critique, in his *Chance and Necessity* (1972), of teleological thinking and the 'vagueness' of the term teleonomy.

> prevent informed and fruitful discussion. For all its manifest inadequacies, the continental-analytic divide is grounded in the undeniable fact that, sometime around the 1920s, philosophers split into two camps that, in short order, had nothing to say to one another. (Gutting 2005, p. 2)

Bergson and Canguilhem, on Gutting's analysis, in some ways defy categorization according to the three areas of the philosophy of science, since the effects of the logical positivist movement in the Anglophone world and Germany somehow bypassed French philosophy, partly, as Gutting interprets it, because it was 'redirected' through to the social sciences. It is then reconfigured through a line of thinkers certainly indebted to the originating thought of Auguste Comte but offering a renewed epistemological impetus to the study of science in the works of Gaston Bachelard through to Georges Canguilhem. Their approach 'accepted the cognitive priority of science and regarded the domain of lived experience as merely a first approximation to the truth about the world, a truth toward which science moved by revising and even rejecting the concepts of everyday experience' (Gutting 2005, p. 11). The positivist spirit of Canguilhem's work is belied by his embracing of a 'vitalism' that unites the concept of life with that of knowledge. His work, like Bergson's, is a *biophilosophy* in which the approach to science is at once epistemological and ontological.[4]

Daniel Nicholson and John Dupré discuss the value of process philosophy to thinking about the phenomena of life in the light of a significant project exploring the potential of 'process biology.' There, the work of Alfred North Whitehead is acknowledged as a significant point of reference within a long history of thinkers of 'becoming' spanning the works of Heraclitus and Aristotle amongst the ancients, to Leibniz and Hegel, and finally James and Dewey (Nicholson and Dupré 2018, pp. 5–6). In passages remarkably reminiscent of Bergson's process thought (within the context of an overview of process thinking from which Bergson's philosophy is also remarkably absent) Nicholson and Dupré set out their naturalistic metaphysics in which nature is constituted by processes *all the way down*:

[4]See Wolf and Wong in Beistegui et al. 2014. *The Care of Life.*

> Instead of thinking of processes as belonging to things, we should think of things as being derived from processes. This does not mean that things do not exist, even less that thing-concepts cannot be extremely useful or illuminating. What it does imply is that things cannot be regarded as the basic building blocks of reality. What we identify as things are no more than transient patterns of stability in the surrounding flux, temporary eddies in the continuous flow of process. (Nicholson and Dupré 2018, p. 13)

The approach of Nicholson and Dupré here is timely and significant in its acknowledgement of the long history of process thinking that also opens up new possibilities for thinking about and incorporating process concepts at play in contemporary biological research as a continuation of that history of philosophy. As they go on to suggest, it seems that the increasing utility of processualism in biology might be part of a natural development of the sciences more broadly towards dynamic models.[5]

2 Questions of Scale

'Until the death of the man and the destruction of the earth, there is no determinate nexus which in an unqualified sense is either the man or the earth' (Whitehead 1967b, p. 204), Whitehead explains, as he takes us through the formation of what counts as any 'thing' from the perspective of a process philosophy. At the heart of Whitehead's speculative cosmology, the philosophy of organism, is an attempt to give expression to every element of our concrete experience, which means the inclusion of 'everything of which we are conscious, as enjoyed, perceived, willed, or thought' (Whitehead 1978, p. 3). The highly specific terms with which Whitehead opts to articulate the processes, and interacting levels of process, that help to form the world are at once intriguing and alienating. It

[5]In particular, Nicholson and Dupré note the suitability of process rather than substance ontology for thinking about living phenomena, and for grounding specific pieces of empirical research in biology. They emphasise the pace of change towards process thought in the realm of physics, followed a century later by a similar trajectory in biology, speculating that perhaps this shift might simply be a natural development of the sciences more broadly (Nicholson and Dupré 2018, p. 39).

is not my aim here to resurrect his terminology in any great detail, but to suggest that some of Whitehead's core aims and principles point the way towards a rich, multi-layered view of reality that speaks to the different effects of viewing communities of living organisms through lenses, so to speak, of varying strengths of magnification. What Whitehead means by a distinction between microscopic and macroscopic is a difference between the smallest unit of experience, an actual occasion, and the broadest 'givenness of the actual world' (Whitehead 1978, p. 129). Actual occasions give us the barest expression of the ultimate character of reality as process. Their processual character is vector-like; that is, each actual occasion is directed towards something else. There is no further reduction of reality to some ultimate substance, or reference to another reality beyond this one that might instead be substantial. The directionality of actual entities is called a prehension. This term divests the more traditional meaning of 'apprehension' of any sense that it belongs to a consciousness (Whitehead 1967b, p. 234). It is not that an identifiable 'I' apprehends the tree over there. Rather, it is simply the case that there is *a taking account of* (what is *there* from the perspective of what is *here*). A prehension, then, is a process of unifying (Whitehead 1967a, p. 72), giving an actual occasion its quality of connectedness to everything surrounding it. When it comes to our sense that we are, on the contrary, perfectly capable of isolating both our selves and the things that we perceive, it is simply that we have jumped to a macroscopic level of description. What we think of, in our ordinary language, as identifiable things are in fact collections or communities of actual occasions, which we might also refer to as a community of prehensions (if we wish to accentuate their directionality). Such communities are the 'organisms' that serve as the focus of Whitehead's philosophy of organism. What we have, then, are organisms that are collections of processes. Furthermore, each actual occasion, each microscopic process, 'repeats in microcosm what the universe is in macrocosm' (Whitehead 1978, p. 215). In order to maintain our manner of expression in tune with the ultimate nature of process, it makes no sense to speak of any thing as possessing a definition or essence other than its passage to another phase of process, 'for it is inherent in the constitution of the immediate, present actuality that a future will supersede it' (Whitehead 1978, p. 215). The sense of each actual occasion as an unfinished process repeats at every level, such that our sense of the identities of things must be redefined as the repetition of reasonably consistent characters across interacting actual occasions. Only when that perception of consistency across phases ceases (becomes past) can we speak of a thing *being what it is.* Thus identity, essence, and being are functions of the past and they are terms that cannot describe reality in its orientation towards the future (Whitehead 1978, p. 90).

Once we start dealing with questions across the broad array of biological sciences, we are examining the interactions of processes that, in Whitehead's terms, are macroscopic. They are communities of prehensions, collectively referred to as 'organisms.' The specificity of Whitehead's use of the term organism might inevitably become confusing, unless we remember that any thing that we identify, from a mineral formation, to a bacterial cell, an elephant, an ocean, or a planet, counts as an organism. They are all collections of the minimal processes that Whitehead calls actual occasions. Any analysis of conventions and problems in biology might be principally undertaken as a categorization of different kinds and levels of process. It is in this light that the next two chapters, 'Transmutation' and 'Symbiosis,' are composed. In 'Transmutation' I begin with an overview of Charles Darwin's *Origin of Species* in terms of its commencement of a categorial or ideological progression of the meaning of 'species' from the emergence of new lines of evolution from 'incipient species' to ecological and ethological categorizations of dominant, despotic, and eventually invasive species. This is not, however, simply a matter of scientific ideology.[6] Whitehead's interest in the processual nature of reality as demonstrable in the repetition of microscopic processes in the macrocosm invites comparison (successfully or otherwise) with the presentation in ecology of interacting patterns across multiple levels of life. The monitoring of populations deals in the analysis of fluctuating communities of multiple interacting species. The solutions to the challenge of monitoring such unruly objects as living populations will take us to a third aspect of Whitehead's thought: his interest

[6]This refers to Canguilhem's understanding of scientific ideology (Canguilhem 1988); see my discussion of it in Khandker (2014).

in the development of quantum mechanics. In *Science and the Modern World*, Whitehead remarks on the implications of quantum theory for our understanding of the nature of transition from one state to another. According to this theory 'some effects which appear essentially capable of gradual increase or gradual diminution are in reality to be increased or decreased only by certain definite jumps. It is as though you could walk at three miles per hour or four miles per hour, but not at three and a half miles per hour' (Whitehead 1967a, p. 129). Under classical mechanics, certain phenomena that might have been dismissed as self-contradictory or absurd, could, under quantum mechanics, now be incorporated as special features or behaviours given alterations in experimental and observational conventions. I will conclude the chapter, 'Transmutation,' with thoughts on some of the applications of certain methodological conventions across the bounds of quantum and biological theory, to underline key difficulties that come with the monitoring of fluctuating and mobile populations, thereby problematizing the labelling of different species as either despotic or invasive.

In the succeeding chapter, 'Symbiosis,' I follow further implications of this albeit relatively brief reference in Whitehead to quantum mechanics in Lynn Margulis' research into the phenomenon of Symbiogenesis. I begin with a return to the text of *Science and the Modern World*, to revisit Whitehead's study of the complex transition from eighteenth- to nineteenth-century philosophy and science. Central to his study is the distinction between a positivist trajectory towards physical reductionism and a 'Romantic reaction' against mechanistic models of life. Whitehead's identification of his own philosophy of organism with the Romantic resistance to mechanism highlights a persisting obstacle to the unification of the physical and biological sciences, guided by the question: 'What is life?' A leap forward of several decades brings us to Lynn Margulis' 'maverick' theory of symbiosis as a significant vehicle of evolutionary development. First, the focus in research into symbiosis on micro-organisms (both long-extinct and extant) offers a radically more-than-human perspective on the story of life on this planet. Whatever we understand life to be, it is at a minimum definable as both autopoietic and as capable of precipitating a cascade of increasingly complex evolutionary innovations that ought to provoke a change in perspective on our macrofaunal occupation of a planet that, if measured in terms of biomass, by rights belongs to other creatures. Second, Whitehead's interest in the epochal leaps favoured from a certain quantum perspective, resonates with the altered perspective on evolutionary change offered by symbiotic theory. The gradualism suggested by Darwinism united with genetics is problematized by the suggestion that some evolutionary innovations occur in much larger leaps of symbiotic union. Furthermore, Margulis' exploration of Serial Endosymbiosis Theory would necessarily entail the use and development of techniques for the detection of processes already completed, evidence of which might be lost in pre-history or obscured by the overlaying of more complex biological functions over time. The methods suggested for this research are essentially methods for the recovery of 'lost time.' In the final section of the chapter, I turn to Margulis' collaboration with James Lovelock in the development of the Gaia hypothesis. I take Gaia to be an attempt to think fundamental processes, or organisms, in Whitehead's sense, up to the level of the earth as organism, understood here again in Whitehead's sense. The chemical model of homoeostasis serves as a mediating position between the mechanism that Whitehead deplores and the philosophy of organism that he favours, whilst also opening up several problems and questions about the precariousness of ecological balance.

3 Methodologies

In the next two chapters of this book, I return to the writings of Henri Bergson for his formulation of certain methodological (and simultaneously ontological) questions that ought to guide the analysis of the living world. Before I summarize the aims of these chapters, and in order to draw closer to the sense of a philosophy of biology, or biophilosophy, explored in this book, I want to turn to the work of biologist Edward O. Wilson, in particular his short book, *Biophilia*. There, Wilson reflects on the methods of the biologist in terms of the kinds of connections we forge between ourselves and other living organisms. We have an 'innate tendency to focus on life and lifelike processes' (Wilson 1984,

p. 1) attributable, from a sociobiological standpoint, to the development of *Homo sapiens* in its preferred, most beneficial, environment. The concerns of more recent developments in biology reflect and enhance this 'inner direction' of biophilia (Wilson 1984, p. 2). We are drawn to complex life-like things as the echo of an innate hunter (the converse of which is the fear of snakes and spiders, creatures that are potential threats). We are drawn to certain landscapes. Habitat selection peculiar to humans seems to be traceable back to the two million years our species spent living in savannah. Parkland, trees, and hills seem to be designed to replicate this scene, as do even the humblest of gardens, and it appears in our general preferences when exploring landscapes. Wilson argues that this may well be a key feature of biophilia, explaining our tendency to be drawn to such landscapes, and to find comfort in them (Wilson 1984, p. 110). Whilst the popularization of the idea of biophilia is manifested in the promotion of 'green spaces' to enhance mental wellbeing from the perspective of public health, and the (problematic) use of animals for the purposes of therapy, I am interested in Wilson's focus upon the practices of the biologist and what he sees as the principal motivations that drive research into living systems. Wilson's descriptions of his own entomological research (on ants in particular) vividly illustrate, better than any rarefaction into psychological terms, the tendency of biophilia to understand, through a combination of observation and speculation, the living organisms one encounters. For example, Wilson's own observations on the dynamics of a colony of leafcutter ants draws on an intricate analogy between different scales of life, in which a multitude of individual insects might function so seamlessly together that their colony takes on the appearance of a gigantic amoeba:

> Its foraging columns snake out like pseudopods to engulf and shred plants, while their stems pull the green pieces down holes into the fungus gardens. Through a unique step in evolution taken millions of years ago, the ants captured a fungus, incorporated it into the superorganism, and so gained the power to digest leaves. Or perhaps the relation is the other way around: perhaps the fungus captured the ants and employed them as a mobile extension to take leaves into the moist underground chambers. (Wilson 1984, pp. 36–37)

A naïve curiosity, a fascination with a single organism, or in the case above, a superorganism, grows into a desire to learn more, and eventually to propose a theory or explanation that, far from settling the matter, merely generates further questions and begins the search anew. As Wilson continues, 'You start by loving a subject. Birds, probability theory, explosives, stars, differential equations, storm fronts, sign language, swallowtail butterflies' (Wilson 1984, p. 65). The beginnings of all forms of enquiry are thus essentially *philosophical*, if we adhere to an etymology of this word for the 'love of knowledge.' Methodologically, Wilson's reflections on the creative processes that underpin scientific research illuminate further overlaps with the work of the philosopher:

> Science is not just analytic; it is also synthetic. It uses artlike intuition and imagery. In the early stages, individual behaviour can be analysed to the level of genes and neurosensory cells, whereupon the phenomena have indeed been mechanically reduced. In the synthetic phase, though, even the most elementary activity of these biological units creates rich and subtle patterns at the level of organism and society. (Wilson 1984, p. 54)

The key, for Wilson, is the productive deployment of analogies as a means of creating unforeseen links between otherwise unrelated phenomena. We return, then, to Bergson's call for the creation of 'fluid concepts,' though it is not the aim of the present book to simply 'enact' or solve this persisting problem of the presentation of temporal phenomena. A return to Bergson's work, by way of Wilson's reflections on biophilia, entails not just a questioning of what life *is*, but an exhortation to us to change the ways in which we encounter and study life, such that we encourage a concern with *how we know* life. Epistemology recapitulates ontology, if you will. In other words, an exploration of some of the methods that have been explored by certain other thinkers in the histories of philosophy, science, and art, to present, conceptually and visually, processes that otherwise elude capture, aid us towards that shift in focus and emphasis that Bergson refers to as philosophical intuition.

In the chapter 'Metamorphosis,' I juxtapose the works of a number of thinkers, commencing with Johann Wolfgang von Goethe, whose collected notes on the developmental processes of plant growth serve as an early example of formalism, founded on an idea of the leaf as protean form for all other configurations of the plant from seed to fruit. The successive developments of some idea of the primacy of form, perhaps best expressed in the theory that ontogeny recapitulates phylogeny, give us an alternative, if contentious, image of the reiterative relation between the microscopic and the macroscopic. My survey of Goethe's botanical writings concludes with the suggestion that his theory of 'metamorphosis' begins to give way to what he terms the 'spiral tendency,' the latter idea suggesting a primacy of process over form. This burgeoning tendency within Goethe's notes readily compares with Bergson's theory of an *élan vital* grounded in the simple observation that our most immediate evidence for the primacy of change is available to us in the inspection of the fluctuating intensities of attention or concentration that defines our own consciousness. Yet our habit of thinking everything from the 'manifold' of our sensory experience to the metaphysical concept of time itself as a succession of discrete units keeps us focused on a version of reality that is as if viewed through the lens of a cinematograph. In the remainder of the chapter I trace the path that started with Goethe's emphasis on the value of morphological study towards D'Arcy Wentworth Thompson's defence of morphology as a route towards the unification of the study of life's processes in biology with physics. His use of drawings, diagrams, and mathematical models demonstrates the means available to us to conceptualize the 'action of forces' that govern processes such as embryogenesis. I conclude my survey of this trajectory towards 'biophysical' explanation in the practice of contemporary British artist, Gemma Anderson, whose attempts to visualize developmental processes in a Paul Klee-inspired project called 'Isomorphogenesis' takes the unification of science and art not in the direction of an increasingly rarefied positivism, but towards an accentuation on different methodologies for thinking process. The purpose of the visualization is to encourage a kind of 'triangulation' between the thinker, the artist, and the object. The artwork, drawing, or diagram is not a representation of a finished concept; it is instead an integral third orientation towards the living process that is being studied.

My aim in the ensuing chapter, 'Reminiscence,' is to expand on this idea of the use of analogies, models, or proxies as a way of exploring Bergson's disagreement with Albert Einstein on the plural nature of time as postulated in the theory of Special Relativity. I start with the problem of interpretation posed by Bergson's use of diagrams in his 1896 book, *Matter and Memory*. The images of the circuit and the cone are meant to serve as explanatory vehicles for the interlocking processes of memory, perception, and action. However, the cone diagram in particular can act as a barrier to a greater understanding of the relation between memory and action if its demarcation of reality into virtual and actual is reinforced by a spatial demarcation between them in the diagram. My aim is to clarify the relation between virtual (memory) and actual (action) through a comparison between Bergson's diagrams and the use of 'embedding diagrams' in physics that are employed to explain the 'anatomy' of black holes. Towards this aim, I engage with Bergson's summary of his reading and objections to the theory of Special Relativity in *Duration and Simultaneity*. His debate with Einstein centres on the relativization of time achieved in the critique of Newton's concepts of space, time, and their relation to the velocity of light. What Bergson sees as the physicist's position as privileged observer would never be recognized by Einstein, though it is thought that Bergson's critique was enough to prevent the award of a Nobel Prize to Einstein in recognition of his theory of Special Relativity.[7] I follow Bergson's criticisms of the concept of plural times in order to clarify his position on the significance of our individual experiences of time and the fluctuations of concentration that indicate to us the processes of consciousness. Our experiences of our own *lived time* indicate something about the nature of time in general. Presenting this processual quality of experience poses a special challenge to the formulation of concepts and diagrams that, being inherently static things, will inevitably fall short of their requirement to indicate change. To address this problem, I turn to the expansion of Special Relativity into the theory of General Relativity, a central part of which is the prediction of the gravitational effects of collapsed stars, that form black holes, on their surroundings. Here, I return to the use of diagrams

[7]For an extended discussion of this debate, see Jimena Canales (2016).

to explain such effects, for their approach to the presentation of multi-dimensional objects (objects in four or more dimensions). I argue that a productive comparison between the conventions of presenting four-dimensional effects in two-dimensional forms in this area of relativistic physics helps to clarify the meaning of Bergson's diagrams that present processes such as the co-creation of memories and perception. A 'reconciliation' between Bergson and Einstein, through the more contemporary work of physicist, Kip Thorne, might be effectuated in our contemporary understanding of their debate through a focus on the methodological affinities between their respective disciplines.[8]

4 Parts and Wholes

In a letter to the journal, *Philosophy of Science*, in 1934, Norbert Wiener, notable for his development of cybernetic theory,[9] comments upon an article published there by J. B. S. Haldane on 'Quantum Mechanics as a Basis for Philosophy' in 1934. Wiener remarks on the way in which this proposal by Haldane might not also point to the basis of quantum mechanics in the thought of Leibniz (Wiener 1934, p. 479). He finds in the antithetical yet synthesized relation in quantum physics between Newton's particle physics and Huyghens' wave theory a certain affinity between Leibniz's presentation of the interrelations between monads. Leibniz's monads which 'mirror the universe, but do not integrate this activity of mirroring into a self-conscious consciousness' (Wiener 1934,

[8]Both chapters draw on the work of artists within traditional and new visual media, but my explorations here are introductory and partial. There was not room within this book to survey what are themselves substantial areas of research in the use of images, analogies, and diagrams in art, philosophy, and science. The chapters here formed part of what was initially a separate strand of research into the 'entanglements' of drawing and thinking in my writing and pedagogy. The diagrams, drawings, some rudimentary attempts at animation, and time-lapse photography involved primarily had value within a 'live' context. The processes that they were intended to enact lose their efficacy if reproduced as still images, which is why they are not contained in this work. The process of writing this book was thus one of triangulation, moving from scientific object and artistic product to philosophical concept. I am especially grateful to Louise Pallister, whose artworks and willingness to discuss her ideas and methods served as the initial provocation for these methodological chapters that form the core of this book.

[9]See Hayles (1999) for an overview of the significance of the Macy Conferences on Cybernetics.

p. 480) present an image of the universe in which microcosm exists in parallel to macrocosm. Similarly, in quantum theory:

> the electron or similar particle is a microcosm. It carries with it a set of proper coordinates embracing the entire universe. In terms of these coordinates we may express everything happening everywhere and 'everywhen,' but the picture we thus obtain is blurred because of our ignoration of the proper coordinates pertaining to all other particles, and we have no unique way, as far as physics has gone at present, of translating statements concerning the microcosm into statements concerning the macrocosm. (Wiener 1934, p. 480)

Haldane, Wiener observes, compares this behaviour and feature of electrons to living organisms. What might be interpreted as randomness in living processes could be more usefully compared to the feature of quantum indeterminacy, likening the ability of living organisms to make choices in their actions based on representations of possible future outcomes to 'the ability of an electron to cross a potential barrier because of a lower potential on the other side of that barrier' (Wiener 1934, p. 480).

Quantum physicist, Walter Elsasser, in a complementary gesture, cites Norbert Wiener's innovations in cybernetics as contributing to great advances in our understanding of life through 'the close analogy that exists between the structure and principles of operation of the nervous system on the one hand and those of certain electrical networks employing feedback devices on the other'[10] (Elsasser 1951, p. 316). I take up this complex of theoretical links in the next chapter, 'Plasticity.' There, the aim is to return to some of the problems raised in the earliest chapters of this book concerning our concepts of ecological balance, and now consider them by way of the thought of process philosopher, Charles Hartshorne, through to the writings of Gregory Bateson. The affinities between a process perspective on life, particularly Hartshorne's recourse to a certain panpsychist vision of the microcosm/macrocosm

[10] Notably, Elsasser's own attempts to establish a quantum physical basis for biology make reference to both Whitehead and Bergson; his references to Bergson's contention that memories are not stored in the brain can be found in Elsasser's books, *The Physical Foundations of Biology* and *Reflections on a Theory of Organisms.*

relation, and a cyberneticist emphasis on de-anthropomorphized processes of communication between living beings, points us towards an 'ecological' understanding of mind that, as Whitehead says of prehensions, repeats in microcosm what the universe is in macrocosm. The key to this alternative expanded view of mind is its challenge, to integrate into, and to enhance the total system. Biophilosophy, we might say, incorporates a biophiliac perspective and the fusion of biology and philosophy as a productive way to think lived time, such that we can re-orient ourselves towards and for our environment. As Wilson reminds us:

> The prevailing myths concerning our predatory actions toward each other and the environment are obsolete, unreliable, and destructive. The more the mind is fathomed in its own right, as an organ of survival, the greater will be the reverence for life for purely rational reasons. (Wilson 1984, p. 140)

In closing this book with some thoughts from William James, I want to emphasize the importance of the need, ever-recurring, it seems, to take heed of James' perspective on the value of philosophy in the face of the increasingly elevated status of scientific discourse and the popular use of scientific explanation in everyday language. As James notes, the sciences tend to be viewed as operating at the cutting edge of knowledge, whilst philosophy stands still. To this, he responds that, in fact, modern philosophy would seem alien to the likes of Aristotle or Descartes, whereas scientific concepts have remained relatively static to the extent that such early thinkers would feel more at home with them (James 1987, pp. 994–995). In response to the problem of dogmatism in philosophy, James points out that philosophers can, and must, incorporate the various methods of the sciences in order to think about things 'in the most comprehensive possible way,' and thereby avoid falling into dogmatism (James 1987, p. 995). Finally, the orientation of philosophy, if taken from James' pragmatist perspective, towards the concrete and the practical through an engagement with the results of scientific analysis holds the key to overcoming objections to philosophy as 'out of touch' with real life. One such form of engagement might involve reflections on the concept of 'extinction,' no longer simply as an integral and inevitable part of evolution by natural selection but also as the real experience of death as a loss. Canguilhem summarizes the contemplation of death that, whilst it 'opens up avenues, frees up spaces and clears the way for novel life forms,' it also shows that 'this opening is illusory, for one day the bell will toll for today's survivors as well' (Canguilhem and Delaporte, p. 89). This recognition of the 'essential precariousness' of life surely forces us to acknowledge its particular value (Canguilhem and Delaporte, p. 90). Whilst we are in the midst of environmental, and more specifically, climate crisis, which is as much a problem of recognizing it as such as it is a problem of the marked impacts of climate change on biotic communities across the globe (e.g., Bangladesh, Australia, and coastal areas of America), debates about whether we call our age 'The Anthropocene' or not (reiterating the age-old debate about whether we are a part of nature or apart from it) do not change the pace of species decline and extinctions that are occurring globally. Thinking differently about the other species with which we share this planet, that, as Lovelock's Gaia hypothesis indicates, is exceptional in its capacity to sustain life, is more vital than ever. As we saw in both the unveiled images of the shadow of a black hole and the extinction of the Bramble Cay melomys, they inspired awe in quite distinct ways, and they spoke of what human beings can accomplish when they act together, and what they can do if they *fail* to act together. This book is written in the light, and hope, of our inquisitive spirit towards the far reaches of the cosmos *and* the deepest recesses of the microcosmos. Interest in both is deeply intertwined and integral to fostering a concern for life at all levels in processes that will continue far beyond the reach of our own lifetimes.

References

Bergson, Henri. 1992. *The Creative Mind: An Introduction to Metaphysics*, trans. Mabelle L. Andison. New York: Citadel Press.

Bichat, Xavier. 1800. *Physiological Researches on Life and Death*, trans. F. Gold, with notes by F. Magendie. Trans. George Hayward. In *Significant Contributions to the History of Psychology 1750–1920*, ed. Daniel N. Robinson. 1978. Washington D.C.: University Publications of America.

Canales, Jimena. 2016. *The Physicist and the Philosopher: Einstein, Bergson, and the Debate that Changed Our Understanding of Time*. Princeton, NJ: Princeton University Press.
Canguilhem, Georges. 1988. *Ideology and Rationality in the Life Sciences*, trans. Arthur Goldhammer. Cambridge, MA: The MIT Press.
Canguilhem, Georges, and François Delaporte. 2000. *A Vital Rationalist Selected Writings from Georges Canguilhem*. New York: Zone Books.
Collaborat, Event Horizon Telescope. 2019. First M87 Event Horizon Telescope Results. I. The Shadow of the Supermassive Black Hole. *Astrophysical Journal Letters* 875 (1): L1. https://doi.org/10.3847/2041-8213/ab0ec7.
De Beistegui, Miguel, Giuseppe Bianco, and Marjorie Gracieuse. 2014. *The Care of Life: Transdisciplinary Perspectives in Bioethics and Biopolitics*. London: Rowman & Littlefield International.
Elsasser, Walter M. 1951. Quantum Mechanics, Amplifying Processes, and Living Matter. *Philosophy of Science* 18 (4): 300–326. https://doi.org/10.1086/287166.
Fulton, Graham R. 2017. The Bramble Cay Melomys: The First Mammalian Extinction Due to Human-Induced Climate Change. *Pacific Conservation Biology* 23: 1–3.
Gutting, Gary (ed.). 2005. *Continental Philosophy of Science*. Oxford: Blackwell.
Haldane, J.B.S. 1934. Quantum Mechanics as a Basis for Philosophy. *Philosophy of Science* 1 (1): 78–98. https://doi.org/10.1086/286307.
Hayles, N.Katherine. 1999. *How We Became Posthuman: Virtual Bodies in Cybernetics, Literature and Informatics*. Chicago: University of Chicago Press.
James, William. 1987. *Writings, 1902–1910*, vol. 38. New York: Library of America.
Khandker, Wahida. 2014. *Philosophy, Animality and the Life Sciences*. Edinburgh: Edinburgh University Press.
Margulis, Lynn, and Dorion Sagan. 1995. *What is life?*. London: Weidenfeld and Nicolson.
Mayr, Ernst. 1988. *Toward a New Philosophy of Biology: Observations of an Evolutionist*. Cambridge, MA: Harvard University Press.
Monod, Jacques. 1972. *Chance and Necessity: An Essay on the Natural Philosophy of Modern Biology*, trans. Austryn Wainhouse. London: Collins.
Nicholson, Daniel J., and John Dupré, eds. 2018. *Everything Flows: Towards a Processual Philosophy of Biology*. Oxford: Oxford University Press.
Watson, James. 2016. Bring Climate Change Back from the Future. *Nature News* 534 (7608): 437. https://doi.org/10.1038/534437a.

Whitehead, Alfred North. 1967a. *Science and the Modern World.* New York: The Free Press.
Whitehead, Alfred North. 1967b. *Adventures of Ideas.* New York: The Free Press.
Whitehead, Alfred North. 1978. *Process and Reality: An Essay in Cosmology*, ed. David Ray Griffin and Donald W. Sherburne. New York: The Free Press.
Wiener, Norbert. 1934. Quantum Mechanics, Haldane, and Leibniz. *Philosophy of Science* 1 (4): 479–482.
Wilson, Edward O. 1984. *Biophilia.* Cambridge, MA: Harvard University Press.
Woinarski, John C.Z., Stephen T. Garnett, Sarah M. Legge, and David B. Lindenmayer. 2017. The Contribution of Policy, Law, Management, Research, and Advocacy Failings to the Recent Extinctions of Three Australian Vertebrate Species. *Conservation Biology* 31 (1): 13–23.

2
Transmutation

The Noisy Miner (*Manorina melanocephala*) is a despot. That is to say, this common native of the Eastern and South-Eastern regions of Australia (including Tasmania), an avian 'honeyeater' (belonging to the family, *Meliphagidae*) has acquired the status of a 'despotic' species, due to its particularly aggressive defence of its territory, and the resulting impact on smaller competing birds. The Noisy Miner's despotic status is perhaps most identified with its tendency to form 'mobs' of around twenty or more individuals, who chase away both potential predators and perceived 'conspecific' competitors (other species of birds) for territory or resources.[1] It dominates territory to such an extent that, despite being a protected species under Australian law, some 'control' (culling) of the species has been discussed as a potential means to provide some of the declining smaller bird species a chance to flourish. A brief survey of the

[1] See Kennedy et al. (2009) for an example of the assessment of the purpose and possible advantages to the Noisy Miners of their mobbing behaviour. Mobbing has proved effective in the exclusion of other avian species from a defended territory. Furthermore, their distinctive call, after which they are named, proves to be an effective tool for the mobilization of fellow individuals, serving to 'maximize' the effects and advantages of possessing such a call (Kennedy et al. 2009, pp. 488–489).

W. Khandker, *Process Metaphysics and Mutative Life*,
Palgrave Perspectives on Process Philosophy,
https://doi.org/10.1007/978-3-030-43048-1_2

ethological and ecological literature on this bird reveals a gradual shift in its status that we might characterize as a 'derogation' of the species. An early description of the Noisy Miner by John Gould in his 1865 *Handbook to the Birds of Australia*, then identified it as the 'Garrulous honeyeater' (*Myzantha garrula*), and points merely to an irritating presence on walks through the Australian forest:

> In disposition it is restless; inquisitive, bold, and noisy, and frequently performs the most grotesque actions, spreading out the wings and tail, hanging from the branches in every possible variety of position, and keeping up all the time an incessant babbling: were this only momentary or for a short time, their droll attitudes and singular note would be rather amusing than otherwise; but when they follow you through the entire forest, leaping and flying from branch to branch, they become very troublesome and annoying. (Gould 1865, p. 575)

We then see, a century later, accounts that range from relatively benign descriptions of a successful, dominant, urban adaptor (Dow 1979; Arnold 2000; Lowry et al. 2011), to decidedly less favourable portrayals of a 'hyperaggressive' bird, a 'reverse keystone species,' a 'despot,' and a potentially invasive species if it should succeed in migrating beyond its indigenous territories (MacNally and Horrocks 2002; Parsons et al. 2006; Montague-Drake et al. 2011; MacNally et al. 2012; Maron et al. 2013).

Beneath the ecological study of the place of any particular species within an ecosystem, operates a range of assumptions about the modes of interaction of individuals, populations, environments, and the processes of adaptation at a genetic level. In this chapter, I want to consider the shifts in species status from success to threat, in terms of the development of a hierarchy of species roles commencing with Charles Darwin's *The Origin of Species*, the relationship between the two key 'macroscopic' processes of heritable variation and environmental change, and later developments in the study of the relationship between these macroscopic (genetic-evolutionary) and 'microscopic' (atomic-level) processes. I will consider the latter relation with reference to the interventions made by theorists in biology and physics interested in the possibilities of a 'quantum biology' or, at the very least, some potential applications of quantum dynamic models to our understanding of biological processes, including the ecological concern with the dynamics of populations. Although we see Darwin attempt to demonstrate the self-shaping dynamic of evolution, later iterations of this in contemporary conservation biology, as exemplified in the case of the 'despotic' Noisy Miner, assume for themselves the directing hand of artificial selection (manifested in 'ecosystem management') in order to determine the fates of different species. Darwin himself only tentatively rejects any idea of final causes or purposes from evolution, but the debate recurs in modern biology about the drivers and direction of evolution. Contemporary approaches to ecosystem management continue to make teleological assumptions about the roles of different species and equally debatable judgments about what constitutes a self-contained, or containable, ecosystem. By increments, we then start to venture into discourses on the purposes of conservation, its practices of annexing land and culling populous species.[2]

1 Mutability as Ontology

Charles Darwin, it is worth remembering, was quite aware of how distinctive and potentially unpalatable his ideas were, comparing the early controversy provoked by Charles Lyell's theory of gradual geological change to his own idea of natural selection, in the hope that it might eventually come to be widely accepted:

> as modern geology has almost banished such views as the excavation of a great valley by a single diluvial wave, so will natural selection, if it be a true principle, banish the belief of the continued creation of new organic beings, or of any great and sudden modification in their structure. (Darwin 1996, p. 79)

[2]Indeed, the continuity of such practices with the persistence of colonialist violence towards indigenous *human* populations.

Thinking beyond the personal dynamics between Darwin and Lyell[3] an important discussion of the nature of species distinctions arises in the transition from Lyell's work, *The Principles of Geology* (c. 1835) and Darwin's *Origin of Species* (1859). For the purposes of this chapter, I want to focus on two key aspects of *The Principles*: first, the way in which Lyell situates his own work against the grain of an ancient heritage of cosmogonical explanation; second, the denial of transmutation from one species into another. This will serve both as an introduction to Darwin's main thesis of evolution by natural selection and its implications for our understanding of the nature of species distinctions, and some reflections on the contemporary reach of Darwinism viewed from a Whiteheadian, as well as a more broadly philosophical, perspective; namely, the problems of ontological process at microscopic and macroscopic scales, and the distinctions between natural and unnatural, or indigenous and invasive, which are themselves dynamic categories or mobile boundaries at work in the intertwined histories of philosophy and the sciences.

Charles Lyell's lengthy historical survey of the roots of geology reveals what he surmises to be an erroneous yet continuing assumption that there was some distinction between ancient causes of geological change and contemporary processes:

> in other words, they supposed that the causes by which the crust of the earth, and its habitable surface, have been modified at remote periods, were almost entirely distinct from the operations by which the surface and crust of the planet are now undergoing a gradual change. (Lyell 1835, p. xi)

Commencing with a range of ancient cosmogonies spanning Eastern and Western thought, including Egyptian ideas and their continuation into ancient Greek philosophy (Plato, Aristotle, Stoicism) and medieval thought, Lyell points to the predominance of a 'catastrophist' view of the creation of the world, typically either by *Ecpyrosis* (conflagration) or

[3] See Ernst Mayr's (1972) summary of the Lyell-Darwin relationship and why the Darwinian revolution was slow to take hold.

Cataclysm ('diluvial' theories).[4] The story that Lyell constructs from his long 'sketch' is of the gradual emergence of insights into a quite different picture of geological change. Ancient causes of geological change are, it is revealed, ongoing. This is Lyell's 'uniformitarianism.'[5] The specific configuration of Lyell's theory, in which geological processes are uniform and cyclical, has particular consequences for how we think about the age of the Earth, the possibility of the transmutation of species (if considered in relation to the speed and nature of geological change), and the effects of human activity on geology. The challenge to the ancient cosmogonical theories is, most notably, of course, a challenge to Creationist under-estimations of the age of the earth. For Lyell, ancient thinkers were clearly venturing into the territory of what would become modern science, but 'the ancient history of the globe was to them a sealed book, and, although written in characters of the most striking and imposing kind, they were unconscious even of its existence' (Lyell 1835, p. 27). Lyell's diagnosis of the ills of the ensuing period of history in the West that frustrated the significant development of geological research, centred on the problem of the interpretation of fossil remains and the possibility of framing any interpretation in terms of the 'Noachian deluge' (Lyell 1835, p. 34). The discovery and analysis of the fossil record, piecemeal over time, and the insights of a few notable theorists who rejected the idea that fossils were simply spontaneously-occurring mineral formations, identifying them instead as the mineralized imprints of once-living organisms, brings with it the intertwining of processes of geological and organic change.[6]

This speculative link between geology and life will also be significant for later studies from both a philosophical perspective and a scientific one. Take, for example, Bergson's discussion of the emergence of early

[4]Lyell, Book I, p. 12. Cf. the persistence of mythological tropes in the 'first philosophers,' such as the image of a founding or generative fire in the fragments of Heraclitus.

[5]A number of commentators note the problematic nature of Lyell's image of an enlightened advancement of human knowledge; see, in particular, Gould (1987).

[6]Lyell identifies a 'favourite dogma' that the fossils of marine animals were not, in reality, of animal origin, grounded in a generalized disbelief that life could have existed before the formation of existing mountain ranges. He refers to the example of Steno, who 'had dissected a shark recently taken from the Mediterranean, and had demonstrated that its teeth and bones were identical with many fossils found in Tuscany' (Lyell 1835, p. 39).

forms of life in *Creative Evolution.* There, he contemplates rudimentary organisms that must have at first proceeded in close tandem with physico-chemical mechanisms from which they would have been barely discernible, such that 'it is hard to say whether they are still physical and chemical or whether they are already vital. Life had to enter thus into the habits of inert matter, in order to draw it little by little, magnetized, as it were, to another track' (Bergson 1998, p. 99). Bergson's speculative moment is echoed somewhat and explored further in more recent research into the problematic taxonomical division of life into five kingdoms (Plantae, Animalia, Fungi, Protista, and Monera). Carl Woese et al. (1990), for example, cite Ernst Haeckel's 1866 attack on the division of life along vegetative and animal lines as barriers to further insights into the evolutionary history of unicellular life. In short, the five-kingdom division breaks, in terms of meaningfulness, at a molecular level since the eukaryote-prokaryote distinction relies upon the presence of characteristics in eukaryotic cells, and merely the absence of those same characteristics in the prokaryotes (Woese et al. 1990, p. 4577). It is by virtue of developments in research at a molecular level, with the ability to identify properties of life previously indiscernible, that three 'domains,' above the level of kingdoms, are proposed: Eucarya, Bacteria, and Archaea (Woese et al. 1990, p. 4579). Accordingly, an altered picture of the emergence of the earliest forms of life is possible:

> The fossil record indicates that photosynthetic eubacteria (and by inference, therefore, archaebacteria and possibly eukaryotes) were already in existence 3-4 billion years ago, so that the evolutionary events that transformed the ancestor common to all life into the individual ancestors of each of the three major groups must have occurred over a relatively short time span early in the planet's history. Both the relatively rapid pace of, as well as the profound changes associated with, this early evolutionary transition argue that this universal ancestor was a simpler, more rudimentary entity than the individual ancestors that spawned the three groups (and their descendants). (Woese et al. 1990, p. 4577)[7]

[7]See also Burns et al. (2004) for a discussion of the earliest evidence available to us in the form of the microfossils known as stromatolites. They point to the rather precarious status of

If we return to Lyell, who is seemingly setting the scene for Darwin's theory of evolution by natural selection, he at first rejects as fantastical the idea of the transmutation of species from simple to complex if the latter means 'more perfect,' and if the efficient cause of change is environmental influence combined with use and disuse in the Lamarckian sense. 'Displacement' not 'transmutation' is the primary effector of changes in populations of species in any given habitat. Lyell's rationale for opting for the former explanation appeals to evidence of the suitability of different species for particular environments, such that

> Even if we thought it possible that the palm or the elephant, which now flourish in equatorial regions, could ever learn to bear the variable seasons of our temperate zone, or the rigours of an arctic winter, we should, with no less confidence, affirm, that they must perish before they had time to become habituated to such new circumstances. (Lyell 1832, p. 173)

At stake is the relationship between two processes: first, the rooting and reproduction of species in climates favourable to their characteristics and, second, the altering pace of geological change. If one were to speculate on how the gradual transmutations might occur, we could, in the manner of a *reductio ad absurdum*, demonstrate just how improbable this process must be: 'We pass at once to the last grand step in the progressive scheme, whereby orang-outang, having been already evolved out of a monad, is made slowly to attain the attributes and dignity of man' (Lyell 1832, p. 14).

In order to dismiss this thought further, Lyell cites the ancient idea of the gradual degeneration of creation from an original perfect state. The gradual process of improvement and complication from simple to complex beings is simply a reversal of this now redundant ancient idea (Lyell 1832, pp. 11–12). Faced with a paucity of evidence, Lyell is demonstrating that

> the only inductively responsible way to develop causal hypotheses about the pre-historic past is to find real causes, observed repeatedly, that are

---

such fossils as evidence, reliant upon the comparison of the visible physical features of modern stromatolites with those of potential ancient examples.

> 'competent, under different modifications, to the production of a great multitude of effects' at least analogous to those effects constituted by the data of paleontology and natural history. (Lennox 2005, p. 88)

Thus, whilst evidence of geological influence on organisms is clear (and an influence that is often fatal, as earthquakes, volcanic eruptions, and tsunamis so clearly show us), the inverse organic influence on geology is minimal. Taking the example of the effects of the 'success' (in the evolutionary sense) of the human species, the extent of human influence is limited to the introduction or replacement of other species, such as the dispersion of plants beyond their original borders or the introduction of *Rattus rattus*, the European rat (and consequent effects on indigenous species), to the 'New World':

> we may regard the involuntary agency of man as strictly analogous to that of inferior animals. Like them we unconsciously contribute to extend or limit the geographical range and numbers of certain species, in obedience to general rules in the economy of nature, which are for the most part beyond our control. (Lyell 1832, p. 122)

It is not, Lyell argues' an 'exclusive prerogative' of human beings to 'conquer' land and to defend that land forcibly, since every other species possesses this impetus resulting in comparable observable phenomena of the domination or even 'extirpation' of some species by others (Lyell 1832, p. 156). Furthermore, not only is human influence simply exploiting a niche or role that would otherwise be occupied by a range of other species, but geological processes have proven to be operating on such a vast timescale that the relatively recent appearance of *Homo sapiens* cannot (yet) be cited as a significant agent of change:

> If all the nations of the earth should attempt to quarry away the lava which flowed during one eruption from the Icelandic volcanoes in 1783 and the two following years, and should attempt to consign it to the deepest abysses of the ocean, wherein it might approach most nearly to the profundities from which it rose in the volcanic vent, they might toil for thousands of years before their task was accomplished. Yet the matter borne down by the Ganges and Burrampooter, in a single year, probably

> exceeds, in weight and volume, the mass of Icelandic lava produced by that great eruption.[8] (Lyell 1832, pp. 207–208)

Lyell's early reflections on the variety and distribution of species firmly rejects any idea of transmutation in favour of the appearance of diverse forms and the effects of migration and encroachment of different species upon one another. The dynamic proper to life demands that a species must succeed or be supplanted by stronger populations, or risk extinction, itself a natural consequence of this fundamental process of life. Lyell will, of course, concede to Darwin's revision of the hypothesis of transmutation, as I will outline in the next section.

2 Darwin and Transmutation

Recall the two levels of process being considered here: first, the rooting and reproduction of species in climates favourable to their characteristics and, second, the altering pace of geological change. Darwin will incorporate a third: the internal and heritable process of adaptation.[9] With Darwin, we find that mutability by insensible degrees, rather than sudden acts of creation, is the primary fact of living organisms and of their encompassing living systems. We are dealing less with the 'origin' of species (some ideal form or repository of potentiality) than with the transition from 'incipient' to established, identifiable species, that 'in order to become in any degree permanent, necessarily have to struggle with the other inhabitants of the country' (Darwin 1996, p. 46). The changing 'conditions of life' (environmental factors) and the mutability of organic bodies favours those individuals that can make the best use of available resources. This is the essential dynamic of natural selection. Furthermore,

[8] Cf. the adoption of this stance for the purposes of biocentric environmentalism. For Taylor (1981), our awareness of our relative insignificance should be humbling and force us to adopt a stance of respect towards a broader ecosystem that would thrive just as well without us, but without which we would perish.

[9] Followed by Hugo De Vries (who contributed to the popularization of Mendelian inheritance), to become the process of pangenesis or genetic mutation. See De Vries (1910).

mutability is not just 'improvement;' it waxes and wanes with changes in the environment.

There are, of course, important caveats to this, illustrated in Darwin's incorporation of certain cultural prejudices against seemingly less developed (and less well evolved) populations of both humans and non-humans in the southern hemisphere, such as those inhabiting Australasia. At this point, it is important to acknowledge the extent to which the theory of natural selection is flexible enough to allow for a non-hierarchical view of species. Darwin is primarily interested in a phenomenon that is in fact quite ubiquitous:

> How have all those exquisite adaptations of one part of the organisation to another part, and to the conditions of life, and of one distinct organic being to another being, been perfected? We see these beautiful co-adaptations most plainly in the woodpecker and misseltoe; and only a little less plainly in the humblest parasite which clings to the hairs of a quadruped or feathers of a bird; in the structure of the beetle which dives through the water; in the plumed seed which is wafted by the gentlest breeze; in short, we see beautiful adaptations everywhere and in every part of the organic world. (Darwin 1996, p. 51)

In his discussion of the 'law' of acclimatization, we see a subordination of any idea of perfectibility to one of adaptation. Species are not simply 'improving' in accordance with some ideal measure of perfection or power. Rather, they demonstrate the capacity to become acclimatized to certain conditions. For Darwin, this presents us with the difficult task of identifying the precise roles of habit, use and disuse, and the role of 'innate differences' in shaping evolution. Once a species becomes acclimatized or habituated to a certain environment, it attains a certain vulnerability through its gradually developed inflexibility (Darwin 1996, p. 115). Recall that Lyell saw the habituation of species to particular environments (his example of elephants in equatorial regions) as proof against the possibility of adaptation, whilst here Darwin merely seeks to account for a tendency that demonstrates that evolution does not necessarily advance towards greater perfection.

With Darwin, it is the continuing development and selection of 'innate differences' that seems to have priority, also illustrated in the evolution of instincts:

> As modifications of corporeal structure arise from, and are increased by, use or habit, and are diminished or lost by disuse, so I do not doubt it has been with instincts. But I believe that the effects of habit are of quite subordinate importance to the effects of the natural selection of what may be called accidental variations of instincts; - that is of variations produced by the same unknown causes which produce slight deviations of bodily structure. (Darwin 1996, p. 171)

Speaking again to a resistance to hierarchical thinking, Darwin identifies a further law in which 'distinct species present analogous variations; and a variety of one species often assumes some of the characters of an allied species, or it reverts to some of the characters of an early progenitor' (Darwin 1996, p. 130). The phenomenon of what we now call convergent evolution illustrates the emergence of similar characteristics in otherwise unrelated species, or the 'reversion' to characteristics thought lost, once belonging to earlier generations of the same species. Characteristics can recur and will, if proving advantageous or, at the very least, not harmful or prohibitive in any way, persist in succeeding generations. A recurrence of characteristics, whether this is across distinct species or along intergenerational lines, demonstrates the importance of efficiency and a certain frugality in natural processes, though this is a point of contention in a number of Darwin's respondents.[10] The complex and yet efficient process of natural selection is nowhere better illustrated than in the cell-making instinct of the 'hive-bee,' and Darwin marvels at observations amongst mathematicians 'that bees have practically solved a recondite problem, and have made their cells of the proper shape to hold the greatest possible amount of honey, with the least possible consumption of precious wax in their construction' (Darwin 1996, p. 183).

Finally, the complex web of 'beautiful adaptations' is underpinned by the relentless struggle for existence waged between individuals and

[10]Cf. Eimer (1890), De Vries (1910), and later Gould (2000).

species both with one another and in the face of ever-changing environmental conditions:

> It is the doctrine of Malthus applied with manifold force to the whole animal and vegetable kingdoms; for in this case there can be no artificial increase in food, and no prudential restraint in marriage. Although some species may be now increasing, more or less rapidly, in numbers, all cannot do so, for the world would not hold them. (Darwin 1996, p. 54)

This returns us to the fate of 'weaker' individuals and species. Using the example of a relationship between certain ants and aphids, Darwin observes that 'each species tries to take advantage of the weaker bodily structures of others' (Darwin 1996, p. 172). Weakness and thus disadvantage manifests in varying forms at varying levels, as we saw in the case of habituation or acclimatization that can give rise to a vulnerability in a species, such as an unsuitability for either warmer or colder climates. The process of refinement or specialization can cause a characteristic to tip over from an advantageous skill to a fatal disadvantage. Commenting on his own experiments on caterpillars and the successive stages of their hammock-weaving, Darwin observes that if he interrupts what he identifies as six distinct stages of construction, the caterpillar is constrained in its possible responses. In one experiment he notes how a caterpillar at the third stage of hammock-building was put into one in the sixth stage of construction, 'so that much of its work was already done for it,' but 'far from feeling the benefit of this, it was much embarrassed, and, in order to complete its hammock, seemed forced to start from the third stage, where it had left off, and thus tried to complete the already finished work' (Darwin 1996, p. 170).

The instinctive action described here is an innate 'skill' that aids the subsistence of the caterpillar in its environment, but when subjected to an impediment (albeit a somewhat mischievous human one), the impetus to work in specified stages could delay the caterpillar in its construction of its protective home and thereby leave it vulnerable to predation or simply to the elements. The emergence of disadvantage from what was previously a strength is also demonstrable, for Darwin, in a general priority of new species over older ones, since the action over time of competition for resources produces better adaptations or stronger characteristics. As Darwin notes, referring to evidence from the fossil record, 'Rarity… is the precursor to extinction' (Darwin 1996, p. 90). The rarer a species, the more vulnerable it is to extinction, which is simply the final component of the process of natural selection.[11]

More ancient species such as the 'living fossils,' *Ornithorhynchus* and *Lepidosiren*, persist only by virtue of relatively tranquil conditions afforded by confined freshwater areas that generate less competition than in the open sea (Darwin 1996, p. 88). Similarly, the marsupials of Australasia, dwellers of the southern hemisphere, are comparatively weak when set alongside more recently evolved 'placental' mammals of the north:

> It may be doubted, for instance, whether the Australian marsupials, which are divided into groups differing but little from each other, and feebly representing, as Mr Waterhouse and others have remarked, our carnivorous, ruminant, and rodent mammals, could successfully compete with these well-pronounced orders. In the Australian mammals, we see the process of diversification in an early and incomplete stage of development. (Darwin 1996, pp. 95–96)

The path to extinction is, then, simply a path towards newer and better-adapted species. However, this does raise a particular difficulty for Darwin, spawning his discussions of the geological record in the latter chapters of *The Origin of Species.* If extinct species are the intermediate steps between extant species (Darwin 1996, p. 102–103), then why is the fossil record not replete with examples of these intermediate species?

---

[11]Consider Darwin's characterization of our negative reactions to extinction as a reflection of our 'profound ignorance' and 'high presumption' such that 'we marvel when we hear of the extinction of an organic being; and as we do not see the cause, we invoke cataclysms to desolate the world, or invent laws on the duration of the forms of life!' (Darwin 1996, p. 61). This echoes Lyell's point in his *Principles of Geology*: '…although we have as yet considered one class only of the causes (the organic) whereby species may become exterminated, yet the continued action of these alone, throughout myriads of future ages, must work an entire change in the state of the organic creation, not merely on the continents and islands, where the power of man is chiefly exerted, but in the great ocean, where his control is almost unknown' (Lyell 1832, pp. 156–157).

Darwin's response is simply that the geological record is incomplete. It is 'a history of the world imperfectly kept, and written in a changing dialect; of this history we possess the last volume alone, relating only to two or three countries' (Darwin 1996, p. 251).

Contrast Darwin's consideration of the debate about what constitutes advancement, and what will come to define some key problems facing the science of taxonomy. Referring to the field of botany, Darwin notes that

> some botanists rank those plants as highest which have every organ, as sepals, petals, stamens and pistils, fully developed in each flower; whereas other botanists, probably with more truth, look at the plants which have their several organs much modified and somewhat reduced in number as being of the highest rank. (Darwin 1861, p. 118)[12]

Advancement might be indicated by the specialization of only certain features in a balance of the relative size of the individual organism, its proximity and reliance upon other members of its species (depending on whether individuals are solitary or social), and its response to the pressures exerted upon it by external conditions. What advantage, Darwin asks, 'would it be to an infusorian animalcule—to an intestinal worm—or even to an earth-worm, to be highly organized?' (Darwin 1861, p. 119) Recalling Lyell's *reductio ad absurdum* that demonstrates the improbability of transmutation, Darwin himself wonders how the first and simplest organisms, perfectly able to subsist and flourish as they were, made their first steps towards 'advancement.' Darwin's response to this is, again, to point to the lack of evidence either way, and to dismiss such speculation as 'baseless and useless' (Darwin 1861, p. 121). Yet teleological explanation is potentially within Darwin's scheme. Harry Smit (2010), for example, admitting first that the idea of Darwinism as teleological is not the standard interpretation of it amongst modern evolutionary biologists, nevertheless argues that a 'rehabilitated teleological

[12]This discussion is featured in comments added by Darwin to an American edition of *The Origin of Species* published in 1861.

explanation' is one that 'leads to more interesting and promising explanations of phenomena and helps us to reconcile evolutionary theory and medicine, psychology, and law' (Smit 2010, p. 358). John Beatty (2016) explores the concept of the 'creativity' of natural selection that is 'best understood in terms of its (1) initiating evolutionary change, and (2) being responsible for the presence of the variation it acts upon, for example by directing the course of variation; as oddly non-Darwinian as both of those points may sound' (Beatty 2016, p. 663). By referring to a creative impetus, so to speak (and he does refer to Bergson here), Beatty seeks to cast light on the processes that give rise to different variations (what he calls 'creativity') as conceptually and temporally distinct from the dynamics of natural selection that effectuate their accumulation or elimination (Beatty 2016, p. 677).

Whatever may be the force that initiates change, the tension between directed and accidental changes was already apparent in Darwin's use of the analogy between artificial and natural selection. As Tabb (2016) notes, the analogy

> while helpful for Darwin's displacement of special creationism, from the outset began to put rhetorical and conceptual constraints upon his thought. *Orchids* works to counter these constraints by weakening the analogy between artificial and natural selection, even as its seductive rhetoric aims to render the theory of selection irresistible and its role in the new science of evolution beyond dispute. (Tabb 2016, p. 12)

3 Contemporary Approaches to Biodiversity

The struggle for existence, the name given to the selective mechanism (deferring the question of any initial creative impetus), dominates more contemporary ecological debates about the place and status of species within their respective ecosystems. Darwin paints a picture of this often-obscured war of all against all in nature:

> We behold the face of nature bright with gladness, we often see superabundance of food; we do not see, or we forget, that the birds which are idly singing round us mostly live on insects or seeds, and are thus constantly destroying life; or we forget how largely these songsters, or their eggs, or their nestlings, are destroyed by birds and beasts of prey; we do not always bear in mind, that though food may be now superabundant, it is not so at all seasons of each recurring year. (Darwin 1996, pp. 52–53)

The very definition of natural selection rests upon the assumption of limited resources, and those species that lose in the competition for resources will become rarer and, eventually, in some cases inevitably, extinct. How does the impetus to manage ecosystems emerge from this discussion? We move from a descriptive view of natural selection to a 'technological' or instrumental view of it. The practice of artificial selection is augmented and carried over into the management of ecosystems based on a definition of the Earth as a limited resource. Over time, if left unchecked, the most 'successful' species may, in fact, exceed the available resources. This has led to concepts of environmental and wildlife management: annexing of national parks, and wildlife control measures, such as culling (for species in excess) and captive breeding (for endangered species). Rosie Ibbotson (2017) notes that, historically, the practice of 'sampling' species (killing a specimen for preservation in a museum) in order to study it (for example, many now extinct or endangered birds in New Zealand) was viewed simply through the lens of 'natural extinction,' whereas there was no acknowledgment of the role played by human beings in the loss of biodiversity. The latter view acknowledging anthropogenic processes, a consequence of later iterations of the concept of biological extinction, is understood in terms of a new impetus to preserve biodiversity through the management of ecosystems. What counts as a 'successful' species is now highly contingent upon the frame of reference for studying such ecosystems. Before even embarking on a discussion of the path of any particular species towards the status of despot or invasive, a wide range of assumptions about the behaviour of ecosystems merits interrogation, as a way of understanding how a species like the ebullient Noisy Miner might be considered an 'interactor' at one time, and an 'invader' at another.

The 'uncertainty' of species categorizations summarized by Hey et al. (2003) and the biases of the observer, insofar as they have an impact on the ways in which particular entities are treated as varieties, incipient species, or definable species, recalls the central insight of quantum mechanics into the behaviour of particles and the effects of the observer on the phenomena observed.[13] Niels Bohr, in his 1933 speculations on the relationship between concepts of 'Light and Life,' pointed to a number of potential routes for aligning the physical and the biological sciences through the properties of light. At the physical level, from the perspective of his own science, Bohr notes that the study of light reveals two complementary aspects, 'the spatial continuity of light propagation [...] and the atomicity of the light effects' (Bohr 1933, p. 422), that must be treated as incompatible if solely viewed from the perspective of classical physics. Similarly, the processes of organic life express characteristics that are not immediately reducible to physical laws. This irreducibility, if acknowledged in the domain of biology, as it is in the domain of quantum physics, should afford us the ability to recognize the limitations of the application of concepts and frames of reference when dealing with the dynamics of living systems:

> in a physical measurement it is never possible to take the interaction between object and measuring instruments directly into account. For the instruments cannot be included in the investigation while they are serving as means of observation. As the concept of general relativity expresses the essential dependence of physical phenomena on the frame of reference used for their coordination in space and time, so does the notion of complementarity serve to symbolise the fundamental limitation, met with in atomic physics, of our ingrained idea of phenomena as existing independently of the means by which they are observed. (Bohr 1933, p. 423)

---

[13]Their survey of key works of the Modern Synthesis illuminates the persistence of a view shaped since Darwin that species display 'special properties' that render them real and objective. They cite Theodosius Dobzhansky's *Genetics and the Origin of Species* for its promotion of modern genetic research in the study of species as 'real genetical and evolving entities,' and Huxley's *The New Systematics* as 'the historical touchstone for modern systematic research programs that see species not just as categories with representatives in museums, but also as dynamic evolving entities that exist independently of human observers and of human-assigned categories' (Hey et al. 2003, pp. 597–598).

In view of this stance, we might problematize the distinction between natural and artificial selection when we consider the nature and roles of human intervention in ecological functioning (e.g., the effects of human activity such as travel, agriculture, etc.) described alongside the dynamics of other species, and the imperative to rectify or rebalance ecosystems. If human activity is not simply, as Lyell would have it, one species amongst many exercising a natural prerogative to secure and defend resources, but rather carries with it an additional imperative to act as 'steward' of nature, then we have already made an enormous leap outside of the realm of scientific analysis into ethics. However, before we can address an ethical stance, a number of other assumptions and adopted frames of reference come into play in understanding the behaviour of ecosystems, the roles of individual species within them, and the effects of human observation and intervention (at times the same thing).

In thinking about the nature of ecological analysis, we might follow Bohr's speculative stance on thinking about organic phenomena through a consideration of different levels of analysis and the interplay between them. Levin (1992) sheds light upon the vast range of perspectives adopted for the purposes of understanding different aspects of an ecosystem (spatial, temporal, organizational), and begins to echo both Bohr and a Whiteheadian account of the world as a system of interlacing prehensions when pointing out that 'every organism is an "observer" of the environment, and life history adaptations such as dispersal and dormancy alter the perceptual scales of the species, and the observed variability' (Levin 1992, p. 1943). Elsasser (1951) explicitly cites an indebtedness to Whitehead in his use of the microscopic-macroscopic relation, with particular reference to 'amplification' as a mode of functioning of living, as well as non-living, systems:

> The physical properties of sufficiently small systems are studied by devices that necessarily involve amplification; only in this way the phenomena of the atomic world can be individually related to the unique effects of the macroscopic world such as the positioning of pointers. (Elsasser 1951, p. 305)

Examples of these 'amplifiers' in the realm of the non-living that Elsasser lists include the Geiger counter, the photographic plate, and the formation of water droplets in a Wilson chamber (Elsasser 1951, p. 305). This role, and more importantly, *limitation*, of such modes of amplification (or mediation) illustrates the affinity that Bohr finds between the physical and the biological. Citing Bohr, Elsasser underlines how the microscopic variables that affect the behaviour of an organism at a macroscopic level elude observation:

> [Bohr] expresses this by saying that if we would try to measure with appropriate precision all those internal variables whose knowledge is required for a unique prediction of its future behavior (for instance by using simultaneously a sufficiently large number of Heisenberg microscopes) then the structure of the organic tissue would be disturbed to such a degree that it would no longer be capable of carrying out its vital functions; in other words the animal would die. According to Bohr this behavior is essential to the operation of all organized matter. He designates it as a generalized complementarity: operation of the vital functions precludes knowledge of the microscopic variables; determination of the microscopic variables precludes biological operation of the living organism. (Elsasser 1951, p. 307)[14]

Elsasser makes two observations in response. First, the incommensurable relation between the microscopic (atomic) and the macroscopic (organic) realms helps us to guard against 'premature, crude and inaccurate interpretations (...), relating macroscopic effects of the organism to specific, localizable individual events in the realm of atomic physics' (Elsasser 1951, p. 308). Second, and as a possible unintended consequence of the

[14]Elsasser expands on this problem in his book, *Reflections on a Theory of Organisms.* The loss in descriptive accuracy that ensues when we try to consider the numerous possible configurations of the elements of organic compounds provides us with a key distinction between physical and biological description and verifiability. Physical laws are assumed to be applicable to infinite sets of 'experiments;' that is, they hold everywhere at all times. Biological principles, on the other hand, can typically only refer to finite sets, because the features are so variable (true individuals are so numerous) that a sufficient amount of data (number of samples) is unavailable. Therefore, the biologist can only deal in rules not laws. Unfathomable complexity refers to this limitation rather than to some unknowable vital principle (Elsasser 1998, p. 41).

first point, this could take us down the path of 'arbitrary vitalistic speculations' unless the *determinacy* of organic life is affirmed and explored further (Elsasser 1951, p. 309).

Let us consider the first point. Stamos (2001) assesses some possibilities for thinking the role of quantum indeterminacy in evolutionary processes at the level of 'point mutations.' Whilst concluding that such a focus is in itself too narrow, Stamos goes on to advocate a wider range of pathways for future research into the links between atomic behaviour and genetic variation. Further, Wolynes (2009) summarizes some experimental work that seems to demonstrate that

> quantum coherence effects play a big role in light energy transport in photosynthetic green sulfur bacteria under physiological conditions. Quantum coherence allows a nonclassical simultaneous exploration of many paths of energy flow through the many chromophores of a light-harvesting complex, thereby significantly increasing the efficiency of the energy capture process, presumably helping the bacteria to survive in low light. (Wolynes 2009, p. 17247)

Such experiments start to push beyond Elsasser's caution against reifying the speculative link between the atomic and the genetic, whilst speculations on the links between the genetic-evolutionary and ecological levels, on the other hand, focus instead on the use of models as a means of incorporating the limitations of 'frames of reference' into an understanding of ecosystems. Taking Elsasser's second point, a certain amount of predictability might be found in a consideration of the fluctuations of populations from a statistical point of view, 'scaling up' from the behaviour of individual organisms to that of identified groups of a species. We saw that Charles Darwin acknowledged the futility of constructing a totalizing narrative of the procession of evolution based on the fragmented evidence of the fossil record, observing that it would be 'as rash in us to dogmatise on the succession of organic beings throughout the world, as it would be for a naturalist to land for five minutes on some one barren point in Australia, and then to discuss the number and range of its productions' (Darwin 1996, p. 48). In the absence of complete and direct evidence of such processes, Levin (1992) as well as

Jørgensen (1990) show how a quantum mechanical 'trade-off' in accuracy of measurement when we move from the level of individual members of a specific population to the broader regularity of the movements of the total population demonstrates the extent to which the design and careful deployment of mathematical modelling has potential benefits:

> Thus we focus attention on larger spatial regions, longer time scales, or statistical ensembles, for which macroscopic statistical behaviors are more regular. This is the principal technique of scientific inquiry: by changing the scale of description, we move from unpredictable, unrepeatable individual cases to collections of cases whose behavior is regular enough to allow generalizations to be made. In so doing, we trade off the loss of detail or heterogeneity within a group for the gain of predictability; we thereby extract and abstract those fine-scale features that have relevance for the phenomena observed on other scales. In physics, this trade-off is well studied, and goes to the heart of the problem of measurement (see, for example, Heisenberg 1932, Planck 1936). At fine scales, quantum mechanical laws must replace classical mechanical laws; laws become statistical in character, dealing only with probabilities of occupancy. (Levin 1992, p. 1947)

Consider Alfred Russel Wallace's comments on 'natural balance' using the example of the now extinct passenger pigeon. He asks how a bird that lays only one or two eggs a season, typically raising only one chick at a time, could become so abundant, especially when we compare it to other birds that produce greater numbers of offspring. The explanation, he continues, must be in the abundance and wide distribution of the pigeon's main food source, whilst the bird itself is 'capable of a very rapid and long-continued flight, so that it can pass without fatigue over the whole of the district it inhabits, and as soon as the supply of food begins to fail in one place is able to discover a fresh feeding-ground' (Darwin and Wallace 1858, p. 55). Conservationists have noted the surprising nature of the decline of this once superabundant species (its population estimated at 3–5 billion in the nineteenth century), its final known example dying at Cincinnati Zoo in 1914 (Halliday 1980, p. 157; Hung et al. 2014, p. 10636). The favourable conditions for the maintenance of such large, roaming flocks, primarily the abundance of its main food source, as recounted by Wallace, were mitigated by its historical population fluctuations (more recently discovered through ancient DNA analysis). Normally, the species experienced 'recurring population fluctuations of large magnitude,' from which it had been able to recover (Hung et al., p. 10637). However, some dramatic fluctuations under certain conditions, that is, a high degree of *stochasticity*, made the species particularly vulnerable to extinction:

> [demonstrated in the hypothesis that] a downward trend in its population size occurred simultaneously with human exploitation in the late 1800s and that the combination of the two triggered its rapid extinction. Once below a minimum threshold population size, the conspicuous roosting and breeding behaviors of this bird could prevent its recovery. (Hung et al. 2014, p. 10640)

The interactions between the genetic composition of individuals within a species deemed, for example, to be not only superabundant, but also 'despotic' or 'invasive' and the ecosystems they occupy are complex. To conclude this chapter, I want to return to the case of the Australian Noisy Miner (*Manorina melanocephala*). This bird typifies what is described as a strongly interacting, or 'keystone,' species, or one that has a disproportionate influence on the collection of other species with which it shares its habitat. The influence of this species is deemed to be disproportionate in relation to the relative size of individuals and its numbers (MacNally et al. 2012, p. 668). In short, if we take the Noisy Miner as an example, given its relatively small size (say, compared to the much larger Australian magpie) offset by its presence in large numbers, and augmented no doubt by its hyper-aggressive behaviour, it exerts considerable pressure on neighbouring species. It is this combination of keystone status and the tendency to assert that dominance, typically in the defence of 'resource-rich territories' using aggressive behaviour, that earns a species its 'despotic' status (MacNally et al. 2012, p. 669).

However, there are considerable problems associated with measuring the effects of a despotic species. Bear in mind that Noisy Miners are a highly visible species precisely because of their success in subsisting in human-dominated habitats (they are, remember, 'urban adaptors'). They are notable for dominating 'patches,' or a somewhat euphemistic way of describing patches of forest left undisturbed between expanses of land developed for human agriculture and habitation. The effects of human activity tend to be excluded from the account of the behaviour of despotic species as an underlying assumption or in historical descriptions of the changes in habitat configuration over time (in the case of the Noisy Miner, this change is set within the timescale of European colonization of Australia). Leaving aside this habitual omission for the moment, and returning to the problem of measurement, we can turn to some potential methods for calculating species densities.

Arguments for the identification of invasive species need to take into account a genetic-level awareness of the 'founder' population; its speed and strength will depend on the genetic health and diversity of that population. To illustrate this, Lambrinos (2004) shows how the European rabbit (*Oryctolagus cuniculus*) was introduced to Australia using only an initial founding population of thirteen individuals. Despite this, the current expanded population shows not degradation in genetic diversity, presumably because of the intentionally rich diversity of the founder population. What this tells us is that a genetically diverse founder will spread more successfully than a genetically narrow one. Thus, the interaction between evolution (genetics) and ecology must be studied in relation to the founding population itself, rather than simply its effects on the biodiversity of the indigenous ecosystem. This study of the founder population might also be undertaken from the perspective of the metabolic rates of interacting species. Brown et al. (2004) explore the possibilities of a 'metabolic theory of ecology' using 'a mechanistic, quantitative, synthetic framework that (1) characterizes the effects of body size and temperature on the metabolism of individual organisms, and (2) characterizes the effects of metabolism of individual organisms on the pools and flows of energy and matter in populations, communities, and ecosystems' (Brown et al. 2004, p. 1786). They note, however, that the overall metabolic rate is complicated by the 'phenotypic plasticity' of some organisms:

> For example, during periods of resource shortages, many organisms are able to lower metabolic rates and resource requirements by reducing activity and entering some form of diapause or torpor. Even these phenotypic

> variations, however, occur within constraints on metabolic rate due to three primary factors: body size, temperature, and stoichiometry. (Brown et al. 2004, p. 1773)

At what point, then, can we say that a species has become 'invasive'? Charles Elton's book *The Ecology of Invasions by Animals and Plants* (originally published in 1958), presents a summary of evidence of the role of invasive species in disturbing ecological balance particularly in small or sparse populations, such as those on remote tracts of land and islands that are typically rendered accessible by the fluctuations of migrating human populations. Elton's study refers to early (laboratory) experiments that indicate the vulnerability of populations with low biodiversity (for example, one predator and one prey species) to drastic fluctuations in the numbers of each species. Recall, also, the case of the passenger pigeon discussed above, the decline of which seemed, ultimately, to be a combination of factors that accentuated its fluctuating population levels (high dependency on specific food sources, and anthropogenic interference in its ability to recover from population fall). Rather optimistically, Elton argues that more complex mixtures of populations facilitate greater overall stability, since species that might otherwise become 'too successful' are subjected to a broader range of checks on their expansion. A greater variety of predator-prey-parasite-disease combinations ensures that overall biodiversity is maintained within a complex system of counterbalances (Elton 2000, pp. 122–123).[15]

The Noisy Miner is characterized as a species with the *potential* to become an invasive one, given its aggressive behaviour and more importantly, it seems, its preference for Eucalyptus forest patches. The proliferation of Eucalyptus plantations worldwide may 'accidentally' lend the

---

[15]Compare the later formulation of the dynamic of island biogeography by Robert MacArthur and Edward O. Wilson (2001): 'By studying clusters of islands, biologists view a simpler microcosm of the seemingly infinite complexity of continental and Oceanic biogeography. Islands offer an additional advantage in being more numerous than continents and oceans. By their very multiplicity, and variation in shape, size, degree of isolation, and ecology, islands provide the necessary replications in natural 'experiments' by which evolutionary hypothesis can be tested' (MacArthur and Wilson 2001, p. 3). And further, a 'basic equilibrium model is developed which postulates immigration rate curves that fall and extinction rate curves that rise with an increase in the number of resident species' (MacArthur and Wilson 2001, p. 66).

bird a route to expansion beyond its indigenous habitats with 'severe consequences' for any native birds it encounters (Maron et al. 2013, p. 1475). More recently, Chris Thomas, in *Inheritors of the Earth* (2017), argues for a rather different understanding of invasive species, whilst attempting to maintain the conservationist's imperative to encourage biodiversity, by considering the total biomass on Earth (overall metabolic turnover), and not simply those species considered 'naturally occurring' or 'wild':

> We cannot reverse time. Instead, we should appreciate changes that are positive as much as we regret any losses. Yes, we have caused the extinction of most of the world's largest land mammals but the American bison is back within the confines of where we allow them to roam. Elephants are doing well in Danum. The grey whale is again migrating in impressive numbers along the western seaboard of North America. Deer and foxes are in our backyards. Nature is fighting back. (Thomas 2017, p. 53)

Perhaps Thomas' position strikes a balance between Lyell's estimation of the impact of human activity on the earth as merely that of one animal amongst many, and the later imperative to track and promote biodiversity amidst increasing levels of environmental degradation. The gradual derogation of species status, as in the case of the Noisy Miner, could be viewed instead as an indication of the capacity of our fellow inhabitants of the Earth to 'fight back' in spite of *Homo sapiens*' tendency to punish other species for their resilience.

References

Arnold, Kathryn E. 2000. Group Mobbing Behaviour and Nest Defence in a Cooperatively Breeding Australian Bird. *Ethology* 106 (5): 385–393. https://doi.org/10.1046/j.1439-0310.2000.00545.x.

Beatty, John. 2016. The Creativity of Natural Selection? Part I: Darwin, Darwinism, and the Mutationists. *Journal of the History of Biology* 49 (4): 659–684. https://doi.org/10.1007/s10739-016-9456-5.

Bergson, Henri. 1998. *Creative Evolution*, trans. Arthur Mitchell. New York: Dover.

Bohr, Niels. 1933. Light and Life. *Nature* 131 (421): 457. https://doi.org/10.1038/131457a0.

Brown, James H., James F. Gillooly, Andrew P. Allen, Van M. Savage, and Geoffrey B. West. 2004. Toward a Metabolic Theory of Ecology. *Ecology* 85 (7): 1771–1789. https://doi.org/10.1890/03-9000.

Burns, Brendan P., Falicia Goh, Michelle Allen, and Brett A. Neilan. 2004. Microbial Diversity of Extant Stromatolites in the Hypersaline Marine Environment of Shark Bay, Australia. *Environmental Microbiology* 6 (10): 1096–1101. https://doi.org/10.1111/j.1462-2920.2004.00651.x.

Darwin, Charles. 1861. *The Origin of Species.* New York: D. Appleton & Company.

Darwin, Charles. 1996. *The Origin of Species.* Oxford: Oxford University Press.

Darwin, Charles, and Alfred Wallace. 1858. On the Tendency of Species to Form Varieties; And on the Perpetuation of Varieties and Species by Natural Means of Selection. *Journal of the Proceedings of the Linnean Society of London: Zoology* 3 (9): 45–62.

De Vries, Hugo. 1910. *Intracellular Pangenesis.* Chicago: Open Court.

Dow, Douglas D. 1979. Agonistic and Spacing Behaviour of the Noisy Miner Manorina Melanocephala, a Communally Breeding Honeyeater. *Ibis* 121 (4): 423–436. https://doi.org/10.1111/j.1474-919X.1979.tb06683.x.

Eimer, Gustave Heinrich Theodor. 1890. *Organic Evolution as the Result of the Inheritance of Acquired Characters According to the Laws of Organic Growth,* trans. J. T. Cunningham. London: Macmillan.

Elsasser, Walter M. 1951. Quantum Mechanics, Amplifying Processes, and Living Matter. *Philosophy of Science* 18 (4): 300–326. https://doi.org/10.1086/287166.

Elsasser, Walter M. 1998. *Reflections on a Theory of Organisms: Holism in Biology.* Baltimore and London: Johns Hopkins University Press.

Elton, Charles S. 2000. *The Ecology of Invasions by Animals and Plants.* Chicago: University of Chicago Press.

Gould, John. 1865. *Handbook to the Birds of Australia,* vol. I. London: Taylor & Francis.

Gould, Stephen Jay. 1987. *Time's Arrow/Time's Cycle.* Cambridge, MA: Harvard University Press.

Gould, Stephen Jay. 2000. *Wonderful Life: The Burgess Shale and the Nature of History.* London: Vintage.

Halliday, Tim R. 1980. The Extinction of the Passenger Pigeon *Ectopistes migratorius* and Its Relevance to Contemporary Conservation. *Biological Conservation* 17 (2): 157–162. https://doi.org/10.1016/0006-3207(80)90046-4.

Hey, Jody, Robin S. Waples, Michael L. Arnold, Roger K. Butlin, and Richard G. Harrison. 2003. Understanding and Confronting Species Uncertainty in Biology and Conservation. *Trends in Ecology & Evolution* 18 (11): 597–603. https://doi.org/10.1016/j.tree.2003.08.014.

Hung, Chih-Ming, Pei-Jen L. Shaner, Robert M. Zink, Wei-Chung Liu, Te-Chin Chu, Wen-San Huang, and Shou-Hsien Li. 2014. Drastic Population Fluctuations Explain the Rapid Extinction of the Passenger Pigeon. *Proceedings of the National Academy of Sciences* 111 (29): 10636–10641. https://doi.org/10.1073/pnas.1401526111.

Ibbotson, Rosie. 2017. De-Extinction and Representation: Perspectives from Art History, Museology, and the Anthropocene. *International Review of Environmental History* 3 (1): 21–42.

Jørgensen, S.E. 1990. Ecosystem Theory, Ecological Buffer Capacity, Uncertainty and Complexity. *Ecological Modelling* 52: 125–133. https://doi.org/10.1016/0304-3800(90)90013-7.

Kennedy, Robert A.W., Christopher S. Evans, and Paul G. McDonald. 2009. Individual Distinctiveness in the Mobbing Call of a Cooperative Bird, the Noisy Miner *Manorina melanocephala*. *Journal of Avian Biology* 40: 481–490.

Lambrinos, John G. 2004. How Interactions Between Ecology and Evolution Influence Contemporary Invasion Dynamics. *Ecology* 85 (8): 2061–2070. https://doi.org/10.1890/03-8013.

Lennox, James G. 2005. Darwin's Methodological Evolution. *Journal of the History of Biology* 38 (1): 85–99. https://doi.org/10.1007/s10739-004-6511-4.

Levin, Simon A. 1992. The Problem of Pattern and Scale in Ecology: The Robert H. MacArthur Award Lecture. *Ecology* 73 (6): 1943–1967. https://doi.org/10.2307/1941447.

Lowry, Hélène, Alan Lill, and Bob B.M. Wong. 2011. Tolerance of Auditory Disturbance by an Avian Urban Adapter, the Noisy Miner. *Ethology* 117 (6): 490–497. https://doi.org/10.1111/j.1439-0310.2011.01902.x.

Lyell, Charles. 1832. *Principles of Geology*, vol. II. London: John Murray.

Lyell, Charles. 1835. *Principles of Geology*. London: John Murray.

MacArthur, Robert H., and Edward O. Wilson. 2001. *The Theory of Island Biogeography Princeton*. Princeton, NJ: Princeton University Press.

MacNally, Ralph, and Gregory Horrocks. 2002. Relative Influences of Patch, Landscape and Historical Factors on Birds in an Australian Fragmented Landscape. *Journal of Biogeography* 29 (3): 395–410. https://doi.org/10.1046/j.1365-2699.2002.00682.x.

MacNally, Ralph, Michiala Bowen Ralph, Alison Howes, Clive A. McAlpine, and Martine Maron. 2012. Despotic, High-Impact Species and the Subcontinental Scale Control of Avian Assemblage Structure. *Ecology* 93 (3): 668–678. https://doi.org/10.1890/10-2340.1.

Maron, Martine, Merilyn J. Grey, Carla P. Catterall, Richard E. Major, Damon L. Oliver, Michael F. Clarke, Richard H. Loyn, Ralph Mac Nally, Ian Davidson, and James R. Thomson. 2013. Avifaunal Disarray Due to a Single Despotic Species. *Diversity and Distributions* 19 (12): 1468–1479. https://doi.org/10.1111/ddi.12128.

Mayr, Ernst. 1972. The Nature of the Darwinian Revolution. *Science* 176 (4038): 981–989.

Montague-Drake, Rebecca M., David B. Lindenmayer, Ross B. Cunningham, and John A. Stein. 2011. A Reverse Keystone Species Affects the Landscape Distribution of Woodland Avifauna: A Case Study Using the Noisy Miner (*Manorina melanocephala*) and Other Australian Birds. *Landscape Ecology* 26 (10): 1383–1394. https://doi.org/10.1007/s10980-011-9665-4.

Parsons, Holly, Richard E. Major, and Kris French. 2006. Species Interactions and Habitat Associations of Birds Inhabiting Urban Areas of Sydney, Australia. *Austral Ecology* 31 (2): 217–227. https://doi.org/10.1111/j.1442-9993.2006.01584.x.

Smit, Harry. 2010. Darwin's Rehabilitation of Teleology Versus Williams' Replacement of Teleology by Natural Selection. *Biological Theory* 5 (4): 357–365. https://doi.org/10.1162/BIOT_a_00058.

Stamos, David N. 2001. Quantum Indeterminism and Evolutionary Biology. *Philosophy of Science* 68 (2): 164–184.

Tabb, Kathryn. 2016. Darwin at Orchis Bank: Selection After the Origin. *Studies in History and Philosophy of Science Part C: Studies in History and Philosophy of Biological and Biomedical Sciences* 55: 11–20. https://doi.org/10.1016/j.shpsc.2015.11.008.

Taylor, Paul. 1981. The Ethics of Respect for Nature. *Environmental Ethics* 3 (3): 197–218.

Thomas, Chris D. 2017. *Inheritors of the Earth*. London: Allen Lane/Penguin Books.

Woese, Carl R., Otto Kandler, and Mark L. Wheelis. 1990. Towards a Natural System of Organisms: Proposal for the Domains Archaea, Bacteria, and Eucarya. *Proceedings of the National Academy of Sciences* 87 (12): 4576–4579. https://doi.org/10.1073/pnas.87.12.4576.

Wolynes, Peter G. 2009. Some Quantum Weirdness in Physiology. *Proceedings of the National Academy of Sciences* 106 (41): 17247–17248. https://doi.org/10.1073/pnas.0909421106.

3
Symbiosis

In *Science and the Modern World*, Whitehead situates his process philosophical project against the backdrop of the development of the scientific method. The shifting modes of scientific observation result in successive revisions of the size and scale of what count as the fundamental units of matter. Invoking William Wordsworth's often-quoted phrase, 'We murder to dissect,' Whitehead underscores the fundamental, and still unresolved, question of the physical basis of reality, one that is required to encompass both non-living and living things. What is it about living organisms that distinguishes them from the rest of matter, and has yet both emerged from and exists in continuity with that same matter that fails to define them completely? From Berkeley, Whitehead takes the broad set of relations between things (over there) unified in one mind (here and now) and replaces a Berkleian unity of mind with 'a process of prehensive unification' (Whitehead 1967, p. 69). The term *prehension* signifies that process of connecting that precedes cognition and is intended to accentuate its sense as an 'uncognitive apprehension.' This description of the rudiments of experience of all living and non-living

W. Khandker, *Process Metaphysics and Mutative Life*,
Palgrave Perspectives on Process Philosophy,
https://doi.org/10.1007/978-3-030-43048-1_3

matter avoids the problem of simple location, incorporating the multitude of perspectives that are so powerfully conveyed in the philosophies of Leibniz and Spinoza. As Whitehead notes:

> In the analogy with Spinoza, his one substance is for me the one underlying activity of realisation individualising itself in an interlocked plurality of modes. Thus, concrete fact is process. Its primary analysis is into underlying activity of prehension, and into realised prehensive events. Each event is an individual matter of fact issuing from an individualisation of the substrate activity. But individualisation does not mean substantial independence. (Whitehead 1967, p. 70)

The problem of viewing nature in terms of a distinction between mechanism and organism is encapsulated for Whitehead in the discord between the eighteenth-century scientific outlook and our real psychological experiences. Furthermore, our real experiences, composed of myriad prehensions, attest to some sort of organic unity, but a scientific scheme can only account for parts that are essentially 'indifferent' to one another. It does not provide, Whitehead continues, 'any elementary trace of the organic unity of a whole, from which the organic unities of electrons, protons, molecules, and living bodies can emerge. According to that scheme, there is no reason in the nature of things why portions of material should have any physical relations to each other' (Whitehead 1967, p. 73). Whitehead's proposed response to this problem is the 'doctrine of organism,' that tries to resist a recourse to either one of two unproductive standpoints. First, the doctrine is to be distinguished from any kind of vitalism, the latter being only a compromise that, ultimately, allows for some sort of mechanism except as 'partially mitigated' in living bodies (Whitehead 1967, p. 79). On that view, any answer concerning the nature of the connection between life and matter is merely deferred. Second, the doctrine of organism opposes a materialism in which objects are only 'very abstract entities, the products of logical discernment' (Whitehead 1967, p. 79). The unification of both aspects of material reality is easily accessible to us if we reflect on the relation between our own minds and bodies. For example, mental states enter into the plan of the whole organism and thus modify every part of that organism all the way down to the level of electrons. This principle of modification forms the cornerstone of the doctrine of organism. Thus, according to an *organic mechanism* '...the molecules may blindly run in accordance with the general laws, but the molecules differ in their intrinsic characters according to the general organic plans of the situations in which they find themselves' (Whitehead 1967, p. 80).

What Whitehead then finds in the poetry of William Wordsworth, is a response to the problem of the incorporation of psychological experiences into the sensory basis of 'discovery' that underpins scientific analysis:

> It is the brooding presence of the hills which haunts him. His theme is nature *in solido*, that is to say, he dwells on that mysterious presence of surrounding things, which imposes itself on any separate element that we set up as an individual for its own sake. He always grasps the whole of nature as involved in the tonality of the particular instance. That is why he laughs with the daffodils, and finds in the primrose thoughts 'too deep for tears.' (Whitehead 1967, p. 83)

Whitehead locates in Wordsworth's 'feeling for nature' an expression of the prehensive unification, in that simplest of terms, of things over there with a mind over here, with all elements of this process immersed in and altered by one another. This is not a mere fall into subjectivism, but rather a provisional realism in which 'every spatio-temporal standpoint mirrors the world' (Whitehead 1967, p. 91).

By the time Whitehead turns to his survey of the sciences in the nineteenth century, his analysis inevitably looks at the significance of Darwin's *Origin of Species* in which the doctrine of evolution is opposed to blind materialism. It offers a view of nature, first, as a process of co-adaptation between the material environment and living organism s and, second, as the overarching 'creativeness' or plasticity of the environment (Whitehead 1967, p. 111). If evolution only described portions of matter, then nothing would evolve, since any bit of matter could be said to have the potential to generate life. On the contrary, if evolution implies a theory of organisms *all the way down*, then we must be able to identify some unique way in which living matter functions that sets it apart from all other collections of physical material. The doctrine of evolution

> cries aloud for a conception of organism as fundamental for nature. It also requires an underlying activity—a substantial activity—expressing itself in individual embodiments, and evolving in achievements of organism. The organism is a unit of emergent value, a real fusion of the characters of eternal objects, emerging for its own sake. (Whitehead 1967, p. 107)

Thus, Whitehead gives us some metaphysical insight into the organic interconnectedness of different levels of process. Prehensive unification is an expression of the way in which 'everything is everywhere at all times' (Whitehead 1967, p. 91), but how does his discussion of the opposition between a 'blindly running' materialism and a doctrine of organism contribute to our understanding of the intertwined development of living and non-living processes?

This chapter seeks an answer to this problem in a brief survey of the concept of *symbiosis*, as both a biological category and a theoretical tool for analysing the relations between different levels of process, cutting across time (geological, evolutionary) and the geographical organization of ecosystems. In order to articulate further the *concrete* configurations of such multiple levels, the first section of this chapter will be devoted to Lynn Margulis' work on the phenomenon of symbiogenesis, or the process of evolutionary change that proceeds not just through the gradual change of genetic mutation, but by processes that are common in microbial versions of replication and reproduction involving the incorporation of parts and characteristics to form new organic wholes. Margulis traces lost or obscured microbial heritages based on piecemeal present configurations in the fossil record and within features and tendencies of our own animal bodies. The methodologies at play in Margulis' research into Serial Endosymbiotic Theory (SET) provide the biological and metaphysical framework for conceptualizing processes *all the way down*, not simply based on precise genetic sequencing, but also on matching functions and processes across taxonomical kingdoms. Analogies, homologies, and the identification of negative indicators such as 'chemical fossils' can be ontologically significant, not just epistemological tools, if they provide evidence of structural 'repetitions' at different levels of process. In the second section of the chapter, I will turn to some of the main features of Margulis' collaboration with James Lovelock's on the Gaia hypothesis. Noting the growing disparities between the life and earth sciences, Lovelock proposes a science of geo-physiology (Gaia) to elaborate upon Schrödinger's reading of the application of the laws of thermodynamics to the basic processes of living systems. That is, Schrödinger's model of life as a process directed towards the reduction of entropy can be applied more generally at a planetary level to identify the possibility of extra-terrestrial life (Lovelock 2016, p. viii). Gaia theory emerges as a means of detecting the conditions for the emergence of life on any planet, as well as forming an explanatory basis for the seemingly exceptional conditions conducive to life on Earth. The Gaia hypothesis had already proven to be controversial on its own terms before discussions of climate change had come to the fore, but I will end this chapter with some reflections on the limitations of this model, considered from a process perspective, for thinking through contemporary environmental challenges such as biodiversity loss and geological change attributable to anthropogenic climate change.

1 Symbiogenesis

In his Foreword to the 1942 edition of Ernst Mayr's *Systematics and the Origin of Species*, another key figure of the Modern Synthesis, Theodosius Dobzhansky, comments upon the increasing diversification of the field of biology, such that specialists in one sub-field, say, mammology, have become increasingly unable to converse with specialists in other sub-fields, such as entomology. The benefits of a greater accumulation of factual knowledge are offset by a loss in the ability to exchange results of research resulting in a failure to augment scientific knowledge. Contributions to the Modern Synthesis were a consolidation of an increasing trend towards the institution of general principles that might re-open flows of knowledge across the multitude of biological sub-disciplines. Ernst Mayr's assessment of some of the specific problems presented to the ever-diversifying biological sciences in accounting for different degrees and paces of change in living organisms begins with what Mayr sees as the keystone of evolutionary theory, that is, a system of classification that reconciles the genetic and environmental factors affecting the evolution of species:

> It is true that the change from the static species concept of Linnaeus to the dynamic species concept of the modern systematist has not entirely escaped the attention of progressive students of genetics and evolution. However, the whole significance of the polytypic species, of the phenomena of geographic variation, of the differences between geographic and other forms of isolation are by no means as widely appreciated among students of evolution and even among taxonomists as they deserve. (Mayr 1999, p. xi)

Some of the principles that Mayr discusses, illustrated with examples from his own specialism, ornithology, elaborate upon the interrelationships between temporal and spatial features of evolutionary change: that is, gradual changes over time and the processes of speciation according to geographical distribution. The particular challenge of classifying distinct species is the identification of true divergences from a common ancestor, precipitated through competition, use, or disuse, as opposed to instances of convergent evolution (the much-used example of the eye, even across kingdoms), or even, as Mayr illustrates, examples of 'loss convergences' of certain characteristics that arise in response to the loss of other characteristics. For example, earlier classifications have identified flying and flightless birds as two separate groups. Later analysis has suggested that their ancestry is more intermixed, since some flightless birds have developed features that have arisen as the result of their loss of flight, and not because they began as distinct species. Similarly, parasitic bees are evolved from 'industrious' bees; but their parasitism brings about characteristics such as bright colouring that might ordinarily allow for separate categorization (Grutte, cited in Mayr 1999, p. 278). Within the professional practice of taxonomy, Mayr talks about the habits of 'Splitters' and 'Lumpers,' where the more minute enumeration of differences by Splitters speaks to the phenomenon of microevolution occurring over the course of relatively short time spans, and the more 'realist' Lumpers operate within higher categorizations (e.g., genera, families, kingdoms) that entail the longer timescales of macroevolution (Mayr 1999, p. 291).

If we turn to the work of biologist, Lynn Margulis, we can see how Serial Endosymbiosis Theory engages in a reconciliation of the two timespans represented in the processes of micro- and macroevolution. Margulis' theory affirms that symbiogenesis and the more widely accepted means of evolutionary change, direct filiation, are not mutually exclusive; rather symbiogenesis is in itself a major vehicle of evolutionary innovation, posing particular challenges to taxonomy in its affirmation of connections between species through unexpected common ancestry and a disruption of animal/plant/bacteria distinctions through the identification of the microbial heritage of seemingly animal characteristics (Margulis 1993, p. 214). Symbiogenesis indicates *micro*evolutionary processes of change, the possibility of which are demonstrable in contemporary experiments, and these in turn give us a window onto macroevolutionary changes, the origins of which are otherwise indiscernible. The fossil record of the pre-Cambrian eras is increasingly revealing the microbial antecedents of many functions of animal cells, but to start to think about these origins, a change of perspective and of scale is required: moving from the shifting patterns of migration of vast flocks of birds and herds of grazing animals and their predators down to the teeming colonies of microbes that live both inside and around these macrofauna. We also move far back across the geological timescale, as Margulis and Sagan's *Microcosmos* proposes, to be presented with a picture of the world as it might have been, spanning the period of the Earth's formation to the Archean Aeon (around 3900 million to 2500 million years ago). The 'big belch' (Margulis 1993, p. 46) signals the process of the release of gases internal to the Earth's core into the atmosphere. This atmosphere is now predominantly anoxic (without oxygen), and it is thus conducive to the emergence of anaerobic microbial life whose colonies form microbial mats and whose presence is even now preserved in the form of hardened mineral structures called stromatolites. Under these conditions, the speculation continues, either life had emerged from the gradual complexification of chemical reactions (often referred to as pangenesis), or life had commenced elsewhere and some remnant of it had been transported to earth on meteorites in a process called 'directed panspermia.'[1]

Regardless of the precise nature of the initial impetus, and it is something that continues to elude replication, the natural history of microbial evolution is equally the story of the origins of life on this planet, starting with the formation of the cell by virtue of certain chemical reactions that cause lipids to form membranes.[2] These self-enclosed membranes containing organelles are to become the minimum units of life:

> To be alive, an entity must first be *autopoietic* – that is, it must actively maintain itself against the mischief of the world. Life responds to disturbance, using matter and energy to stay intact. An organism constantly exchanges its parts, replacing its component chemicals without ever losing its identity. This modulating 'holistic' phenomenon of autopoiesis, of active self-maintenance, is at the basis of all known life; all cells react to external perturbations in order to preserve key aspects of their identity within their boundaries. (Margulis 1993, p. 56)

Here, Margulis and Sagan build on the work of Ilya Prigogine on dissipative structures to explain the possible mechanisms that might have projected life forwards to more complex evolved forms. The concept of dissipative structure, based on observations of inorganic chemical phenomena, seems readily applicable to organic phenomena. It provides a solution to the problem of uniting physical description with the seemingly incommensurate processes of life. As Prigogine and Nicolis explain, the laws of thermodynamics are predicated on the existence of a certain high degree of order in the past that progressively deteriorates over time towards a more chaotic state, or thermodynamic equilibrium (Prigogine and Nicolis 1971, pp. 109–110). This evolving state, in purely physical terms, is opposed by the phenomenon of biological evolution which

[1] Adherents of directed panspermia included Francis Crick. The ideas of pangenesis and panspermia are themselves ancient ideas, at least in terms of human timescales. For a discussion of the differences between the two concepts in Aristotelian thought, see H. De Ley, 'Pangenesis Versus Panspermia: Democritean Notes on Aristotle's Generation of Animals,' in *Hermes* 108. Bd., H. 2 (1980), pp. 129–153.

[2] See Leslie Orgel (1998) on the early origins-of-life experiments attempting to synthesize organic compounds in replications of early Earth chemical-atmospheric conditions.

appears to be advancing to greater degrees of 'coherence' or order. In their article, 'Biological order, structure, and instabilities,' Prigogine and Nicolis attempt to ground an explanation of the increase in biological order within the bounds of existing physical laws, rather than relying on some concept of exceptional functioning. A clue to solving this problem is in observations of 'systems behaving "chaotically" (thermodynamically) in some cases and in a coherent way in others' (Prigogine and Nicolis 1971, p. 110). They cite certain chemical combinations that under specific conditions (suspended in solution, then stirred continuously) start to exhibit structures that diverge from their original form:

> A characteristic feature of chemical instabilities is that their occurrence depends on a minimum level of dissipation and on specific non-linear types of kinetics. The dissipative structure attained subsequently is a really *new state of matter* induced by a flow of free energy under nonequilibrium conditions. In this new state we have a new physical chemistry on a *supermolecular level* while the laws referring to the molecular level remain unchanged and given by the quantum mechanical or classical equations of motion. (Prigogine and Nicolis 1971, p. 143)

Without modification of physical laws, extrapolation of the above processes could, by stages, help to explain the formation of complex organic, self-replicating structures such as RNA, 'the first sentence in the language of nature' (Margulis 1993, p. 57).

Once the leap from the formation of basic organic molecules to autopoietic structures is made, a much larger, and more improbable, leap requires explanation. How do the very small cells that characterize bacterial organisms give rise to the more complex eukaryotic cells that underpin the subsequent emergence of complex animal and plant life? Margulis proposes Serial Endosymbiotic Theory (SET) as a means of explaining the evolutionary leap to eukaryotic cells. The basic underpinning insight of SET is that 'Three classes of organelles cilia, mitochondria, and photosynthetic plastids were once free living bacteria that were acquired symbiotically in a certain sequence by other, different bacteria' (Margulis 1993, p. 1). The components of eukaryotic cells, or organelles, listed here, Margulis argues, evolved a variety of capabilities or innovations in their earlier independent state and, through symbiotic unions, became mitochondria in animal cells and plastids in plant cells (Margulis 1993, p. 4). For example, we can look to the observable similarities between the evolutionary innovations of bacteria and the characteristic operations of modern multicellular organisms, such as DNA recombination, photosynthesis, symbiosis, meiosis, and even 'programmed death' (Margulis and Sagan 1995, p. 131). This suggestion that animal cell organelles were once free-living bacteria points towards a cascading process of developing symbiotic relationships that themselves helped to forge new evolutionary pathways for life. Thus, not only in our evolutionary history, but also in the present functions of our bodies, we can see evidence that the basic effort to delay thermodynamic equilibrium undertaken by the originating bacteria is simply continued at multiple levels within the conglomerated animal body. The identification of these extremely early evolutionary innovations centres on a period in the Earth's history predating the time commonly referred to as the Cambrian period from around 600 million years ago. Within the preceding Proterozoic eon, dating from 2500 million to 600 million years ago and, further, the Archean eon that spans 3900 million to 2500 million years ago, Margulis engages on a necessarily speculative and yet carefully argued defence of symbiogenesis as a major driver of evolutionary change.

SET is built on an affirmation that all organisms require other species in order to complete their biological processes. For example, and despite expectations, it is the case that in a constant environment, single species rarely exclude all others (that is, different species rely upon the availability of other species for food sources, with the exception of cases of cannibalism) (Margulis 1993, p. 171). Furthermore, Margulis underlines that symbioses (using the terms holobiont-biont relationships) are very common, pointing to specific examples such as the cultivation of bioluminescent bacteria in (or by) certain species of fishes, the 'cold illumination' being advantageous to fishes in the attraction of prey or the repulsion of predators, and in modes of courtship (Margulis 1993, p. 173). The symbiotic union of algae and fungi that we see in lichens demonstrates the evolution of a complex interdependency where the resulting organism (lichen) is unlike either individual partner species as it would appear on its own. However, the mechanisms for symbiogenesis are also difficult to discern for the challenges they pose to models of evolutionary change. 'Chloroplast enslavement' is difficult to identify in modern attempts to observe this process; for example, the mere ingestion of potential new organelles can be prematurely attributed to cases in which the ingested microbes can 'remain functionally autotrophic for weeks inside animal tissue' (Margulis 1993, p. 190). That is, what looks like a new symbiosis may simply be a very slow process of digestion that will, ultimately, terminate in the destruction of the ingested microbe.

In *The Logic of Life*, François Jacob reflects on the impact of the invention of new technologies on scientific analysis, noting that even the invention of the microscope was simply an application of abstract theories of light, and the world of 'swarming forms' that it opened up was slow to alter existing conventions of interpretation of the relation between the lives of macro-organisms and microbes. The presence of this hitherto unseen microcosm was even (for Buffon) 'a flagrant insult to the whole living world' (Jacob 1973, pp. 14–15). Letting go of deeply embedded, because eminently useful, distinctions can take inordinate amounts of time. In the case of lichens, whose separated component species look and behave in radically different ways to the resulting symbiotic partnership, the emergence of a new species echoes the emergence of new material characteristics from the combination of inorganic chemicals that give rise to dissipative structures. Margulis notes that the usual terms with which we characterize species such as lichens, or the bacteria that live in our bodies (rather than the bodies that depend on the bacteria), betray an anthropocentric bias, cemented in the use of binomial nomenclature (Linnaeus' system of classification), that prejudices attention towards the most visible member of the symbiotic relationship:

> We people tend to see and name only the most superficial or largest members of the consortium; we then act upon the self-deceptive construction that the consortium is an independent 'individual.' (Margulis 1993, p. 216)

It thus affirms a unidirectional account of evolution moving from the least complex to the most. For Margulis, symbiosis is better represented by bushes 'with anastomosing branches' (Margulis 1993, p. 215).

Erwin Schrödinger, in his short book *What Is Life?*, presents the relatively large size of organic molecules, let alone complex organisms, as a particular problem in the effort to unite atomic physics with biology.[3] As Prigogine and Nicolis observed in their work on dissipative structures, biology appears to be at odds with physics in the application of the second law of thermodynamics. On the way to addressing the application of thermodynamic principles to life, Schrödinger tackles the problem of simply describing organic processes in continuity with inorganic ones.[4] Living things are so big because it takes an enormous number of atoms in combination for them to begin to display any order, and thus for the organism to benefit from the relative stability of physical laws both internally and in their interactions with their external environment (Schrödinger 1992, p. 18).

Our anthropocentric bias is also felt in the tendency to prioritize life that subsists in the atmosphere, whereas life is thought to have evolved (e.g., through emergence of dissipative structures and autopoiesis) at the intersection of land, air, and sea (Margulis and Sagan 1986, p. 65). That

[3]The principal obstacle to this unification, Schrödinger suggests, is essentially a difficulty originating in an ocularcentric prejudice towards scales of phenomena that are immediately available to our own particular organs of sense. Rather, the microscopic atom does not need to be visible to the human eye in order for it to have an impact in the scheme of organic life. On this footing, Schrödinger explains that when we consider the behaviour of atoms, mechanical laws only apply in a *statistical* sense to the motions of sufficiently large numbers of individual atoms, a principle that then allows for the mechanical description of organic functions. Disorder at the individual atomic level gives way to order, or the appearance of regularities, as the magnitude of interacting atoms increases (Schrödinger 1992, p. 10).

[4]In *What Is Life?*, Erwin Schrödinger acknowledges the prevailing limitation upon physics and chemistry in explaining how living organisms originate and function, but ultimately sees 'no reason at all for doubting that they can be accounted for by those sciences' (Schrödinger 1992, p. 4). Pre-dating, and in some respects inspiring, the work that led to the discovery of the structure of DNA by Crick and Watson in 1953, Schrödinger compares the 'material carrier of life' to an 'aperiodic crystal,' a structure at once amenable to physical description yet possessing an extreme degree of complexity (Schrödinger 1992, p. 5). The search for the 'building blocks' of life in this manner takes place against the backdrop of a long history of enquiries and debates about the vehicle of heredity, from Charles Darwin's tacit assumption of an as yet unidentified carrier, to Hugo De Vries' rediscovery of Gregor Mendel's experiments on inheritance in plants. De Vries' own interpretation of Darwin's idea of 'pangenesis' overcomes, he believes, the limitations that he sees in Herbert Spencer postulation of 'physiological units,' Weismann's 'ancestral plasms,' and Nageli's idioplasm (De Vries 1910, pp. 50–59). Instead, the basic thesis of pangenesis is that 'in the cells there are numberless particles which differ from each other, and represent the individual cells, organs, functions and qualities of the whole individual' (De Vries 1910, p. 63).

is, 'the visible world is a late-arriving, overgrown portion of the microcosm, and it functions only because of its well-developed connection with the microcosm's activities' (Margulis and Sagan 1986, p. 66). This revised focus upon our microcosmic origins and continued dependency allows for a renewed programme of research into the forms of life that evolved *before* the Cambrian period. Until the 1950s, it was thought that life had commenced in the Cambrian period, but subsequently, fossil evidence has suggested that life had in fact evolved in the *pre*-Cambrian period. However, the means of finding it would need a focus on 'chemical fossils,' that is, the mere *indications* of organic matter that can be gleaned through an analysis of the chemical composition of ancient rocks. Indications of life can be deduced from the presence of certain chemical concentrations that are unlikely to have been synthesized by ordinary prebiotic processes. Such a renewal of focus has begun to suggest that life may in fact be *almost* as old as the surface of the Earth, that is, around 3400 million years old.[5]

We have seen how phenomena such as the evolution of lichens raise problems in the analysis of the contributing species to any symbiotic union; in particular, the effects of symbiosis can be so dramatic that they obscure the original component species. We have also considered examples of some of the models that have been employed to reflect on symbiotic mechanisms in the absence of directly observable processes. In this oblique perspective on the evolutionary paths of life, SET challenges dominant or orthodox interpretations of Darwinian evolution. Whilst, as Darwin notes, the absence of evidence in the fossil record may be a combination of unfavourable conditions for the preservation of fossils, it also points to the likelihood that the 'missing' evidence never existed in the first place. Symbiosis challenges a wholesale application of direct filiation, which is the traditional mechanism assumed by Darwinian evolution. Margulis notes that intermediate species between cyanobacteria (prokaryotes) and algae (eukaryotes; plants) do not exist in the fossil record, not because they died out, but because they never existed (Margulis 1993, p. 55). In fact, Margulis argues, SET and direct filiation are not mutually exclusive. Cyanobacteria are still ancestral to plants in

[5]See, for example, Walter et al. (1980).

some way: i.e. they are the ancestors of plastids contained within plant (and all eukaryotic) cells, but not wholly ancestral to the plant kingdom (Margulis 1993, p. 55). Direct filiation, more broadly, offers a smooth continuist framework for understanding the interactions between genotype, phenotype, and environment. Not entirely in opposition to this, SET grants that biology is extremely conservative, and therefore that similar characteristics will persist (at higher levels of taxa). Innovative leaps, achieved through symbiotic unions, have persisted; whilst greater flexibility in forms have arisen as 'improvisations' from the basis of these underpinning innovations.

In *What Is Life?*, Margulis and Sagan defend the 'most controversial' component of SET, the heritage of undulipodia ('cell tails and chromosome movers') in ancient spirochete 'invasions' of early eukaryotic cells, by speculating on how the depth of their embedding in modern animal and fungal cells is so great that they have become mere 'genetic shadows of their former selves' (Margulis and Sagan 1995, p. 103). They refer to biologist David C. Smith for his invocation of the image of *a grin without a cat* to describe the relation between this possible ancient heritage and the sparse remaining indicators of the original incursions that have since become complex symbioses (Margulis and Sagan 1995, p. 104). A significant barrier to thinking the continuities between the organelles in eukaryotic cells in plants and animals and their prokaryotic ancestry is their differing abilities to function in aerobic and anaerobic conditions. Margulis highlights how the dependence of animal bodies on oxygen overlays an anaerobic heritage.[6] Whilst we recognize our dependency on an oxygen-rich atmosphere, SET demonstrates a potential process for the

[6]Consider the example of the evolutionary precedence of RNA replication as a means of transmitting and exchanging genetic material amongst primitive bacteria. When inserted into the broader story of the evolution of symbiotic relationships that ultimately leads to the present configurations of animal life as symbionts with once free-living bacteria, the process of DNA replication appears as a relatively recent innovation in the transmission of genetic material. RNA can replicate without DNA, but the inverse is not the case: 'That all cells, given ribose sugar, can make deoxyribose from it suggests that ribose came first. RNA, with ribose, evolved before DNA. DNA sugar metabolism evolved by subtraction of oxygen from RNA sugars. The earliest cells may have been beings which only later evolved DNA systems. Comparing RNA and DNA metabolism is an example of peering into cellular windows for clues to life's most ancient origins' (Margulis and Sagan 1995, p. 64).

acquisition of 'shortcuts' to adaptations to changing atmospheric composition through the incorporation of functions developed by different micro-organisms.

Margulis' work to trace the evolution over time of such an array of functions derived from multiple species focusses on 'semes,' or interacting metabolic pathways (multigenic traits), involving collections of enzymes and the DNA to code for them. It is not just individual mutations, but their collections, that provide selective power, whereas selection of isolated traits (enzymes, RNA molecules, etc.) would be 'absurd.' Semes can become extinct over time, but the end products of those pathways can persist over very long periods. Now, this focus on semes distinguishes Margulis's work from that of Carl Woese.[7] Woese's recategorization of life in terms of a level above that of Kingdoms, the 'Domains' of the Eucarya, Bacteria, and Archaea, relies on a highly specialized form of ribosomal RNA sequencing for their identification that, in Margulis' view, is simply uninformative (Margulis 1993, p. 99). Instead, the attempt to retrace the development of semes, and their eventual extinction, focuses on their remnants in surviving processes in extant organisms, such as the processes of heat-shock and the production of stress proteins. This search for 'the grin of the Cheshire Cat' begins with otherwise familiar characteristics of animals and plants that, from a renewed perspective, can be shown to have developed in an anaerobic environment in archeabacteria (Margulis 1993, p. 130).

One final example to help reconstruct the symbiogenetic development of undulipodia comes in the form of the organism, *Mixotricha paradoxa*. This microbe is made up of what appears to be at least five different kinds of organism, 'the protist trichomonad host; three different prokaryotic surface symbionts – large spirochetes, small spirochetes that produce the anomalous movement, and their associated "anchoring" bacteria,' and a fourth unidentified bacterial endosymbiont (Margulis 1993, p. 197). In addition to this, the *Mixotricha* is itself a symbiont that resides in the intestines of its own host, an Australian termite called *Mastotermes*

[7] As referred to in the previous chapter; see Woese et al. (1990).

darwinensis. Under the conventional terms of direct filiation an explanation for the evolution of undulipodia traces a gradual process of differentiation from prokaryotic to eukaryotic organisms in which features of the eukaryotic cell, the nucleus and the process of mitosis (cell division for the purposes of reproduction), evolved prior to the development of undulipodia. By contrast, according to SET, free-living spirochetes were ingested and incorporated by host bacteria that eventually evolved into the protoctists, or the first cells containing a nucleus (Margulis 1993, p. 274). Thus the story told by the five-part symbiosis that constitute the body of the *Mixotricha* lends a certain credence to the possibility that complex eukaryotic cell structures that make up our own bodies may well have been constructed by similar means. The capacity of bacteria or prokaryotic organisms to reproduce through a free exchange of genes affords them a massively greater capacity for the creation and persistence of new characteristics than that possessed by most plants and animals. As Margulis and Sagan underline, 'the huge numbers of synchronized interconnected cells in the animals and plants compromise our genetic potential,' whilst bacteria lack specificity to such an extent that they can exchange genes with all other bacteria (Margulis and Sagan 1986, p. 88). Simply put, we are too big and too complex to be sensitive to tiny genetic alterations, but our bacterial predecessors would have been small enough and simple enough to accumulate and 'lock in' such changes. Over vast periods of time, these innovations would become essential to the functions of our bodies and yet imperceptibly advantageous in themselves when measured against the rest of the complex mammalian body.

In the next section, I want to propose a continuity between this chapter's opening questions, from a Whiteheadian perspective, on the methods available to us to think about living systems as both distinct from, and continuous with, non-living processes, and Lynn Margulis'collaboration with the chemist James Lovelock in the development of the Gaia hypothesis. We saw that Whitehead's doctrine of organism discerned the 'prehensive unification' of the perspectives of all living and non-living things; it affirmed that 'everything is everywhere at all times' (Whitehead 1967, p. 91). I have discussed some of the ways in which certain developments in biological research might be interpreted as expansions of Whitehead's doctrine, centred particularly on the complicated reconciliation of the physical and the biological. This final section completes the trajectory of the preceding sections with an exploration of Gaia as a potential continuation of the doctrine of organism, and some of the questions it raises about the theoretical adequacy of the Gaia hypothesis in the face of increasing threats to ecological integrity and biodiversity.

2 Gaia

James Lovelock's presents his own reflections on the problematic distinction between living and non-living processes through an image of a stretch of beach exhibiting different features that indicate different physical states. An empty, untouched beach represents the state of inertia or thermodynamic equilibrium that, in practice is never possible due to ever-shifting weather-patterns. The shifting sands that occur due to the effects of the wind and the sea represent the steady state. A sandcastle is a visible indicator of life, whilst the presence of life itself indicates that there must have been a builder or a creator (Lovelock 2016, pp. 30–31). Lovelock asks how it might be possible to make the distinction between sandcastles and merely incidental sand dunes; in other words, is there a way to distinguish between the works of a system that he names 'Gaia' and 'the chance structures of natural forces?' (Lovelock 2016, p. 31). A response to this comes through Boltzmann's characterization of entropy 'as a measure of the probability of a molecular distribution' (Lovelock 2016, p. 31), that is, the ability to distinguish a distribution of molecules as distinct from its background, such that it constitutes a distinct entity. This provides a means of thinking about the Earth's behaviour (its distribution of molecules) as something improbable, exceptional. Lovelock models what the Earth's progress towards a steady state might look like, such that it ends up resembling the conditions on either Venus or Mars. In reality, what we have is something quite different and extraordinary: Gaia, or a concatenation of checks and balances that ultimately ensures that harmful things, such as the despotism of species, are avoided, or that toxic substances are recycled, and that is, on the whole, beneficial to the continuation of life on the planet.

Lovelock cites James Hutton's 1785 comparison of the global cycling of water with the blood circulation of animals, one that sets an early precedent for what would become the Gaia hypothesis (Lovelock 2016, p. xxiv):

> We have since defined Gaia as a complex entity involving the Earth's biosphere, atmosphere, oceans, and soil; the totality constituting a feedback or cybernetic system which seeks an optimal physical and chemical environment for life on this planet. The maintenance of relatively constant conditions by active control may be conveniently described by the term 'homoeostasis.' (Lovelock 2016, p. 10)

Presented in this way, Gaia is a hypothesis put forward to allow for the experimental exploration of the operations and interdependencies of different living processes (Lovelock 2016, pp. 10–11). We might use the hypothesis of the emergence of self-regulation on a global scale to think anew about the interdependency of the biota, the atmosphere, and the geological composition of the Earth. It can help us to think differently about the workings of evolutionary processes on both species and on ecosystems. The internal and external dynamics of speciation entail both a shaping by and shaping of the physical environment. Consider the effects of 'keystone' or despotic and invasive species, such as the Noisy Miner (as in the previous chapter), on other species and habitats; the effects of the (re-) introduction of top predators into a habitat; the effects of industrious species such as the beaver on a landscape; the delicate balance maintained in lakes and rivers, and yet swiftly upset by small temperature fluctuations that precipitate 'algal blooms' that in turn deoxygenate the water with fatal consequences for fish populations. We can also consider the effects of the removal of certain species from the ecosystem through anthropogenic extinctions.

What follows is an overview of some of the key arguments and examples that Lovelock utilizes to support his thesis of the Earth as a unique cybernetic system, in combination with some of Margulis' insights into symbiotic processes. Their combined vision of the Earth as an immense autopoietic system ties the story of the geological formation of this planet to the emergence of microbial life. The story of this co-formation: (i) ties the evolution of macro-organisms to geology, in the form of foundational evolutionary symbioses (endosymbiosis); (ii) continues in extant symbiotic relations (e.g. termites and their wood-digesting gut-bacteria); (iii) is demonstrable in the role of microbes in recycling dead matter for use by the living; and finally, and (iv) indicates a structural analogy between the basic relationship between a single cell and its environment *all the way up* to the global regulation of the surface and atmosphere of the earth to the benefit of life.

As we saw in Lynn Margulis' examination of our collective heritage in bacterial symbioses, the conditions on the Earth shortly after its formation into a relatively stable planet were ideal for the emergence of life. Contrary to our own (mammal-centric) expectations, what constituted the ideal conditions for life included an anoxic, 'reducing' (that is, hydrogen-rich) atmosphere, as opposed to one dominated by oxygen (oxidization being the action that, for example, rusts iron). The microbial mats that are thought to have initiated the dominance of life on this planet would have found the presence of oxygen at today's levels (roughly 20% of atmospheric composition) far too volatile and would have been 'burned out' by its action. Lovelock goes on to demonstrate how it is possible to glean some confirmation of the necessity of a reducing atmosphere for life in the Earth's earliest eras through comparison with our nearest planetary neighbours. Both devoid of life, Venus lacks hydrogen altogether, whilst Mars locks it in water but its oxidized surface is ultimately anathema to life (Lovelock 2016, pp. 16–17). Not only do we have such comparisons to indicate to us the necessary conditions for life, but the Earth's own geological record yields some intriguing insights into atmospheric fluctuations and marks of bacterial generational shifts in banded iron formations found in sites such as the Grand Canyon. The striations of rock signal fluctuations in atmospheric hydrogen and oxygen levels, referred to as the reduction-oxidization or 'redox' potential of the atmosphere, probably regulated by differing rates and kinds of bacterial respiration. Ultimately, the gradual increase in atmospheric oxygen is thought to be attributable to bacterial respiration, setting off an evolutionary arms race to compensate for the harmful effects of oxidization in living bodies. Consider Margulis' references to the ability in the mammalian body to revert to anaerobic respiration when its physical exertion overruns its intake of oxygen. Rather than an adaptation to low-oxygen conditions, this has been read, under the banner of SET, as a memory of the body's non-mammalian heritage.

The combination of emergent dissipative structures and autopoiesis at cellular level, forming the integral whole upon which life in general functions and evolves, commences an ever-broadening process of environmental change. The self-regulating system is not simply the contents of the cell bounded by a secure, if semi-porous, membrane. Its circular process of internal-external exchange of beneficial and wasteful products requires the cell, or collection of interdependent cells, to regulate its immediate environment. We have already seen, in the example of bacterial respiration and the record of its fluctuations in banded iron formations, an example of vital and environmental co-regulation. Lovelock's justification for reading the entire planet as one extremely large self-regulating system is based on interpretations of what are effectively exceptional life-promoting conditions on Earth. If we wish to discern such processes, Lovelock favours a 'black box' method that can reveal features of a cybernetic system not by pulling apart the system itself, but by varying aspects of its environment on which it is able to exert some control. These perturbations ought to be gentle, not dramatic; as alluded to earlier in the works of Niels Bohr and Walter Elsasser, we can see some advantages of this methodological shift in moves away from *post mortem* dissection to the study of living animals in their natural habitats (Lovelock 2016, pp. 53–54).

What might such a method reveal to us in an analysis of the functions of the Earth's atmospheric processes? If we take the example of the composition and regulation of the gases in the atmosphere, nitrogen in its chemical equilibrium state would exist in the form of nitrate ion dissolved in the sea. However, as Lovelock explains, there appears to be an advantage to keeping the bulk of nitrogen in gaseous form; dissolved in the sea it would raise the salinity of the water. In turn, high levels of salt in the oceans would be lethal to life. Instead, the current processes of regulation of salinity ensure the integrity of the cell membrane, the core essential disequilibrium tool for all organisms. In short, we might conclude that oceans are not more saline than they are because certain 'Gaian' mechanisms are in place that act to remove salt (Lovelock 2016,

p. 85). Furthermore, nitrogen gas in the atmosphere helps to build pressure. It also helps to dilute oxygen concentration to safe levels. If the oxygen were to increase to over 25% in the atmosphere, it would fuel a catastrophic increase in forest fires (Lovelock 2016, pp. 72–73). Thus, the regulation of atmospheric gases at their current levels helps 'Gaia' to stave off conditions that would otherwise be fatal for the living creatures it encompasses. Toxicity and pollution are, Lovelock contends, 'natural' to Gaia. Whilst Gaia's maintenance of atmospheric nitrogen levels helps to regulate overall atmospheric pressure (and by extension, weather patterns), suppress increases to harmful levels of atmospheric oxygen, and ensures that ocean salinity is kept to an optimum level, the current scientific consensus on the progression of climate change presents urgent challenges to what, theoretically at least, seems to be an extraordinarily resilient system.

Take the familiar example of the effects of global warming on the rates and levels of melting of the polar ice caps. This gradual decline has in turn been cited as the cause of around 25% of global sea level rises, and, in Iceland, for example, the loss of ice mass is linked to increased volcanic activity with immediate and serious impacts on life.[8] Add to this the reports of an accelerating decline in insect diversity in Europe since the 1950s,[9] including a calculated 76% loss of biomass of flying insects in Germany in one 27-year study,[10] though tempered somewhat by the sensationalized prospect of a global 'Insect armageddon' with calls to undertake more wide-ranging studies of insect populations.[11] Should the effects of receding ice caps and glaciers be supplemented with such a significant loss of insect diversity that it severely impedes the vital cycles of pollination and support, in the form of subsistence, to other animals, the Gaia hypothesis predicts only that some sort of compensatory processes will occur, but not that these compensations will be to the benefit of all species, least of all our own. As Donna Haraway summarizes, there is an efficiency to Gaia that might be mistaken for a certain serendipity.

[8] See von Hippel and Harig (2019).
[9] See Jan Christian Habel et al. (2019).
[10] See Hallmann et al. (2017).
[11] See Pedro Cardoso and Simon R. Leather (2019).

However, 'Gaia does not and could not care about human or other biological beings' intentions or desires or needs, but Gaia puts into question our very existence, we who have provoked its brutal mutation that threatens both human and nonhuman livable presents and futures' (Haraway 2016, p. 44).

Kirchner (2002) and Lenton and Wilkinson (2003) have gone further in debating the verifiability of Gaia theory, contesting the levels of production of negative (stabilizing, life-promoting) and positive (destabilizing, life-limiting) feedback. Lenton and Wilkinson defend the references to the production of negative feedback, without referring to them as exclusively Gaian (as Kirchner claims they are). Kirchner's objection to the coherence and applicability of Gaia theory to real processes centres on the claim that biological systems produce and sustain external environmental conditions that are beneficial to themselves. However, many such processes do not, and the current consensus on the harmful effects of climate change are a good example. Kirchner's criticisms of Gaia lead, then, to a call for further research into the Earth's regulatory systems that are not necessarily so neatly regulated in line with the needs of the planet's current inhabitants (Kirchner 2002, p. 404). Lenton and Wilkinson respond with the affirmation that

> Life is not always going to enhance gross primary productivity, resistance, resilience or any other metric of the system. There will always be examples for and against. What matters is the balance of examples: Do they suggest a tendency in one direction? For example, is there a tendency toward regulatory feedback or not? The more examples we have, the more confident we can be about answering such questions. (Lenton and Wilkinson 2003, p. 2)

We can at least conclude from this exchange that Lovelock's emphasis on Gaia as a provocation to future research still holds, and that, as Latour (2017) observes it may at least provide an image of 'connectivity without holism' (Latour, p. 75).[12]

[12]As Latour argues: 'The scientists [Lovelock] is struggling against believe you can have penetrability of agents on the one hand, and then stick nonetheless to the idea of an agent plus

Whitehead's challenge to think process from prehensions all the way up, as a way of answering the problem of physical reductionism, is echoed in the challenges posed to scientific orthodoxy in the theories of Symbiogenesis and Gaia. On several counts these scientific hypotheses carve out new paths for a reconfigured image of life's fundamental processes. Serial Endosymbiosis Theory promotes a change in attitude towards 'inferior' forms of life; whilst the broader story of microbial evolution applied all the way up to the formation of a global system of regulation contributes to an altered perspective on the interdependencies of Earth's ecosystems. Yet the integrity of this set of interdependencies, and the proposed capacity of the Earth to regulate any imbalances, is problematic, particularly in the face of real environmental degradation.

In the next two chapters, I want to explore how a Bergsonian understanding of life can encourage the generation of methods for thinking anew about our relationships with more-than-human beings and the wider environment that sustains us.

References

Cardoso, Pedro, and Simon R. Leather. 2019. Predicting a Global Insect Apocalypse. *Insect Conservation and Diversity* 12 (4): 263–267. https://doi.org/10.1111/icad.12367.

De Ley, Herman. 1980. Pangenesis Versus Panspermia Democritean Notes on Aristotle's Generation of Animals. *Hermes* 108: 129–153.

De Vries, Hugo. 1910. *Intracellular Pangenesis*. Chicago: Open Court.

an environment, whereas Lovelock frees their science by extending the mechanistic worldview and distributing agencies at every point along the causal chains. Read in this way, Lovelock, far from "fighting reductionism," has unlocked the explanatory power of reductionism. It is just that no organism can be reduced to its own action! To be fully reductionist, you need to follow through the other actions that are complicit in its action. In that sense Lovelock is as far as possible from any "holistic" thought. Engineers cannot be fooled by the Myth of the Machine. Gaia, for Lovelock, could be called No Machine and that's why, of all the metaphors he criticizes, none is damned more relentlessly than Spaceship Earth' (Latour, p. 77). This echoes the physicalism of Prigogine; cf. also D'Arcy Thompson's efforts to develop a framework for biophysical explanation in the next chapter.

Habel, Jan Christian, Michael J. Samways, and Thomas Schmitt. 2019. Mitigating the Precipitous Decline of Terrestrial European Insects: Requirements for a New Strategy. *Biodiversity and Conservation* 28 (6): 1343–1360. https://doi.org/10.1007/s10531-019-01741-8.

Hallmann, Caspar A., Martin Sorg, Eelke Jongejans, Henk Siepel, Nick Hofland, Heinz Schwan, Werner Stenmans, Andreas Mu Lier, Hubert Sumser, Thomas Horren, Dave Goulson, and Hans de Kroon. 2017. More Than 75 Percent Decline over 27 Years in Total Flying Insect Biomass in Protected Areas. *Plos One* 12 (10). http://doi.org/10.1371/journal.pone.0185809.

Haraway, Donna J. 2016. *Staying with the Trouble: Making Kin in the Chthulucene*. Durham: Duke University Press.

Jacob, François. 1973. *The Logic of Life: A History of Heredity*, trans. Betty E. Spillman. New York: Pantheon Books.

Kirchner, James W. 2002. The Gaia Hypothesis: Fact, Theory, and Wishful Thinking. *Climatic Change* 52 (4): 391–408. https://doi.org/10.1023/A:1014237331082.

Latour, Bruno. 2017. Why Gaia Is Not a God of Totality. *Theory, Culture & Society* 34 (2–3): 61–81. https://doi.org/10.1177/263276416652700.

Lenton, Timothy M., and David M. Wilkinson. 2003. Developing the Gaia Theory: A Response to the Criticisms of Kirchner and Volk. *Climatic Change* 58 (1): 1–12. https://doi.org/10.1023/a:1023498212441.

Lovelock, James. 2016. *Gaia*. Oxford: Oxford University Press.

Margulis, Lynn. 1993. *Symbiosis in Cell Evolution: Microbial Communities in the Archean and Proterozoic Eons*. New York: W. H. Freeman and Company.

Margulis, Lynn, and Dorion Sagan. 1986. *Microcosmos: Four Billion Years of Microbial Evolution*. Berkeley: University of California Press.

Margulis, Lynn, and Dorion Sagan. 1995. *What Is Life?*. London: Weidenfeld & Nicolson.

Mayr, Ernst. 1999. *Systematics and the Origin of Species from the Viewpoint of a Zoologist*. Cambridge, MA: Harvard University Press.

Orgel, Leslie E. 1998. The Origin of Life—A Review of Facts and Speculations. *Trends in Biochemical Sciences* 23 (12): 491–495. https://doi.org/10.1016/S0968-0004(98)01300-0.

Prigogine, Ilya, and Gregoire Nicolis. 1971. Biological Order, Structure and Instabilities. *Quarterly Reviews of Biophysics* 4 (2–3): 107–148. https://doi.org/10.1017/S0033583500000615.

Schrödinger, Erwin. 1992. *What Is Life? With Mind and Matter and Autobiographical Sketches*. Cambridge: Cambridge University Press.

von Hippel, Max, and Christopher Harig. 2019. Long-Term and Inter-Annual Mass Changes in the Iceland Ice Cap Determined from GRACE Gravity Using Slepian Functions. *Frontiers in Earth Science* 7: 171.

Walter, M.R., R. Buick, and J.S.R. Dunlop. 1980. Stromatolites 3400–3500 Myr Old from the North Pole Area, Western Australia. *Nature* 284 (5755): 443–445. https://doi.org/10.1038/284443a0.

Whitehead, Alfred North. 1967. *Science and the Modern World.* New York: The Free Press.

Woese, Carl R., Otto Kandler, and Mark L. Wheelis. 1990. Towards a Natural System of Organisms: Proposal for the Domains Archaea, Bacteria, and Eucarya. *Proceedings of the National Academy of Sciences* 87 (12): 4576–4579. https://doi.org/10.1073/pnas.87.12.4576.

4
Metamorphosis

Philosophy, Bergson writes, was born of an awareness of the insufficiency of our faculties of perception. Retelling the progress from the early Presocratic thinkers of Miletus, whose reflections 'on nature' were grounded in some sense of the primacy of the elements (water, fire, air), to the Eleatic school's criticisms of change, Bergson traces a fundamental, and lasting, break between our trust in sensory experience and our conviction (and perhaps a hope) of a pure, abstract, reality that transcends our perception (Bergson 1992, p. 132). However, the connection between matter and thought, percept and concept, remains. Philosophy is always built upon matter and our experience of the world, no matter how dissociated the resulting rarefied abstraction seems from our messy, concrete existence. If our concepts and ideas (to use Kant's demarcations of the understanding and reason) are constructed as a kind of scaffolding for our perceptions, then 'there cannot be *a* philosophy as there is *a* science' (Bergson 1992, p. 133). Each philosopher augments her concrete experiences with concepts, giving rise to a multitude of philosophies each reflecting the arbitrary choices of its author, whilst the scientist proceeds ever forwards in the drive to boil down reality to its most general essences. Art, by contrast again, aims 'to show us, in nature and in the mind, outside of us

W. Khandker, *Process Metaphysics and Mutative Life*,
Palgrave Perspectives on Process Philosophy,
https://doi.org/10.1007/978-3-030-43048-1_4

and within us, things which did not explicitly strike our senses and our consciousness' (Bergson 1992, p. 135). Bergson lauds the achievements of the great artists (citing Turner and Corot) for their ability to make us see in things something that we had not noticed before, but when we see through their art, it is now an experience fundamentally changed. What they attain is an extension of the faculties of perception (Bergson 1992, p. 136). Yet he reserves for philosophy the capacity to refine our perception of things, by concentrating ourselves in an act that is at once 'perceiving and willing' such that our vision of things is expanded (Bergson 1992, p. 134). The separation between the sensory and the rational, personified in the break between the Milesians and the Eleatics, might be reversed in a renewed effort in metaphysics to perceive and think the nature of change. This separation between the aesthetic and the intellectual and, by extension, art and philosophy, turning on the distinctions between percepts and affects, on the one hand, and concepts on the other, is repeated in Deleuze and Guattari's attempt to answer the question: *What is Philosophy?* Raising the artist and the novelist to the status of *seer* or *becomer*, as Bergson only in part suggests by referring to art's capacity to extend perceptual experience, we find that 'the death of the porcupine in Lawrence and the death of the mole in Kafka are almost unbearable acts of the novelist. Sometimes it is necessary to lie down on the earth, like the painter does also, in order to get to the "motif," that is to say, the percept' (Deleuze and Guattari 1994, p. 171).

The artist perceives and feels; the philosopher thinks. This age-old scission is unhelpfully hierarchical (gendered, anthropocentric), and is, ultimately, a barrier to the possibilities for philosophy that Bergson hints at but perhaps closes off in his only rudimentary engagements with art. What if the question of the distinction between philosophy, art, and science was to be explored at the level of concrete experience? This would not be an enquiry towards a grand unification of disciplinary traditions that span millennia, but rather one aimed at exploring the possibilities of 'doing philosophy' from an epistemological perspective, and of communicating philosophical ideas from a pedagogical perspective. In her book, *Drawing as a Way of Knowing in Art and Science*, Gemma Anderson sets her work within the context of an 'educational turn' in contemporary art practice, 'combining strong educational motivations and an interest in sharing artistic process rather than product' (Anderson 2019, p. 9). In a complementary manner, I take as my starting point in this chapter my own use of drawing in order to think through philosophical problems and to convey philosophical ideas to others as an integral part of my pedagogical practice, to question whether drawing and diagram-making can serve as a means of thinking differently about what it means to be a living organism. This is the more 'local' aim of the next two chapters. My interest in drawing here concerns, principally, its activity or enactment, and secondarily, its status or use as a representational form. On both counts, my suggestion will be that the act of drawing diagrams as integral to the explanation of Bergson's theory of consciousness merits further enquiry. Drawing is, in Bergson's sense, part of our enactment of time. We are not 'in' time, in the manner that objects occupy parts of space. We *are* time, unfolding at different speeds. Time, if any 'thing,' is a heterogeneous multiplicity of speeds and slowness (as Deleuze phrases it). Thus, when one draws, one acts and thinks temporally. Each section of this chapter places this set of problems within a broader tradition of integrating art into scientific research.

The first section examines the 'botanical writings' of German Romantic poet, Johann Wolfgang von Goethe, in particular his *Metamorphosis of Plants* (1790), as a significant example of formalism in the history of evolutionary theory and developmental biology. Serving as a backdrop to debates around the claim that 'ontogeny recapitulates phylogeny,' I follow Goethe's assertion of the protean form of the leaf as the basis of all other parts of the plant that plays out a tension between the process of metamorphosis (in line with form) and a 'spiral tendency;' the latter appearing to resist reduction to any particular form, in a movement of growth that might allow us to think of organic development in terms of an ontologically prior Process. In the second section, I turn to Henri Bergson's *Creative Evolution* for his discussion of the methodological shift required for us to truly think and encounter 'life.' The *élan vital*, or 'vital impetus,' proves to be elusive to our habitual methods for encountering our surroundings (and even our own consciousnesses) since our concepts, the tools of human intelligence, are designed to freeze the flow of reality into discrete parts for the purposes of practical action.

This cinematographical mechanism works well enough in its purely practical orientation, but it is inadequate for the apprehension of the processes of life, and it calls for a revision of our methodological approaches to the analysis of life. In the final section of the chapter, I trace the above lineage from Goethe and Bergson through to D'Arcy Thompson's *On Growth and Form.*[1] In this work, Thompson promotes the use of principally mathematical visualizations to illuminate the convergence of developmental forms across very different species, expanding further on Goethe's integration of a visual or spatial understanding of organic growth as essential for biological research. I then conclude the chapter with an artist's response to the problem of presenting living processes, referring to Gemma Anderson's practice of 'Isomorphology,' which seeks to liberate classificatory systems from their conventional usage, and to incorporate drawing as a way of meditating on form and function across species and broader physical lines.

1 Metamorphoses

In Stephen Jay Gould's 1977 analysis of the history of developmental biology, *Ontogeny and Phylogeny,* he focuses on the contention of the nineteenth-century champion of Darwinian evolution, Ernst Haeckel, that 'ontogeny recapitulates phylogeny'. Ontogeny refers to the life history of an individual, both embryonic and post-natal (Gould 1977, p. 483); whilst phylogeny indicates the evolutionary history of a lineage, conventionally depicted as a sequence of successive adult stages (Gould 1977, p. 484). If ontogeny recapitulates phylogeny, then the growth of an individual mammal, from embryo to adulthood, repeats in its development the adult forms of its evolutionary ancestors. The study of ontogeny was used to yield up clues about the relationships between

[1]See Esposito (2014) for an overview of a range of largely negative assessments of Thompson's challenge to Darwinian functionalism, and the seemingly 'maverick' biologist's indebtedness to a broad international network of biologists and philosophers, including Haldane, Whitehead, and Bergson, that provoked his 'third philosophical way between reductionist and vitalist positions' (Esposito 2014, p. 89).

diverse species, setting them together in a succession of developmental stages within evolutionary trees. From the perspective of the process thought discussed so far in this book, that of Alfred North Whitehead, we might say that the relation between ontogeny and phylogeny presupposes the existence of a formal unity between the microscopic and the macroscopic manifestations of time, but reiterated at the larger scales of the individual organism and the series of species deemed to belong within an evolutionary developmental series leading up to the 'most developed' species. Whether this iterative structure works in the particular instance of the theory of recapitulation will be subject to debate, as we will see in this chapter.[2] For now, let us simply consider Haeckel's work for the principles of biological research that it reflects. As Gould explains, Haeckel writes within an era of biological enquiry underpinned by the observations of increasing complexity during the process of ontogeny and the identification of a scale of species ascending from 'lower' to 'higher' forms, such that 'if there is but one path of ascent to man, and if a human embryo must begin in [Lorenz] Oken's 'initial chaos', then the stages of human ontogeny must represent the completed forms of lower organisms. As Oken stated in his colourful metaphor, what are the lower animals but a series of human abortions?' (Gould 1977, p. 36) However, the assumptions guiding this method, one that infers humanity's gradual emergence from its less developed 'animal' states, are inescapably anthropocentric, perhaps almost to the point of absurdity.

Gould summarizes the influence of Romantic thought on the theory of recapitulation in terms of two guiding beliefs. Firstly, it holds to an 'uncompromising developmentalism' in which nature is seen as irreducible to the properties of matter, progressing relentlessly from an initial state of chaos through to human beings (Gould 1977, p 36). That is, human beings are the end or pinnacle of a process of evolution that

[2]Applying this principle more generally, the reiteration of the macroscopic in the microscopic is a central feature of Henri Bergson's theory of consciousness that, in his 1907 book *Creative Evolution*, is shown to echo the process of the evolution of life. The almost genealogical progression of Bergson's writings, commencing with his study of individual consciousness in *Time and Free Will*, to the more expansive study of consciousness (through memory) and ontological time in *Matter and Memory*, to the evolutionary and cosmological perspective of *Creative Evolution* each time takes a step back, so to speak, to broaden the scope of our visual field in order to better discern the origins of each phenomenon we have studied.

encompasses other, less developed, species that exist alongside them. Secondly, the Romantic view incorporates a concept of harmony and unity between nature and its laws, where 'Man is the highest configuration of matter on earth, but we are indissolubly linked to all objects as the goal toward which they strive. Nature and spirit, the inorganic and organic, are one' (Gould 1977, p. 36). In the subsequent theory of recapitulation it would follow from this that the stages of development of a human being from embryo to death recall the adult forms of adjoining older and less complex species. We might refer to these two beliefs as an iteration of the ancient philosophical idea of the Great Chain of Being[3] but let us focus on the idea here that Form is ontologically prior to Process. The significance of formalist theories in the history of biology is articulated in another of Stephen Jay Gould's works in the history of science, *The Structure of Evolutionary Theory.* Cited by Gould as a key figure just preceding the development of theories of transmutation in the nineteenth centuries, from Cuvier and Geoffroy, Lamarck, Darwin and Wallace, to Haeckel, Weismann and Eimer,[4] is the German Romantic poet Johann Wolfgang von Goethe.

Goethe cites the botanist Wenderoth for the choice that confronts us in the study of biological development:

> It depends on whether we wish to pursue the plant in its living metamorphosis as a 'something' capable of existence only *in regulated alteration*, or whether we wish to grasp and retain it as something constant, and therefore dead, in one or several widely separated specific situations. (Goethe 1989, p. 114)

Favouring the former approach, the focus becomes 'the process by which one and the same organ makes its appearance in multifarious forms,' or the *metamorphosis of plants* (Goethe 1989, p. 31). The documentation of this process is broken down into the key stages of the developing plant, from seed to stalk and towards the full formation of leaves, to flower and to fruit (inflorescence and infructescence). The development of the leaf

---

[3]See Arthur Lovejoy's series of lectures, published as *The Great Chain of Being.*

[4]For my discussion of this lineage in relation to the work of Henri Bergson, see Khandker (2014), Chapter 1.

which, in the first and second essays on Metamorphosis, underpins the formal organization of all the other stages, dominates the first epoch of metamorphosis. Once complete, the next epoch is that of the flower but, again, with reference back to the leaf as organizing principle, it appears that what were once leaves organized along the shaft of the stem, have now become organized around a fixed centre to form a calyx (Goethe 1989, p. 42). In this common principle demonstrated in the early stages of plant development, Goethe sees a foreshadowing of 'the natural force by which the inflorescence and infructescence will be effected at a more advanced age' (Goethe 1989, p. 43). Note that there is already a tension between the formalism of this account, and the concept of metamorphosis (and later the spiral tendency) emerging in passages such as this one:

> By repeating here a remark made earlier, that styles and stamens represent the same stage of development, we can further clarify the cause of this alternate expansion and contraction. From seed to fullest development of stem leaves we noted first an expansion; thereupon saw the calyx developing through contraction, the petals through expansion, and the sexual organs again through contraction; and soon we shall become aware of the maximum expansion in the fruit and the maximum concentration in the seed. In these six steps Nature ceaselessly carries on her eternal work of reproducing the plants by means of two sexes. (Goethe 1989, pp. 60–61)

However, Goethe's essay, if considered within the history of developmental biology, is characterized in terms of its 'proto-formalist' inspiration, anchored as it is in the protean form of the leaf:

> When we consider a plant in relation to its vital force, we see this vitality manifesting itself in two ways: first, through *vegetative growth*, by development of stems and leaves; and next, through *reproduction*, which is completed in the formation of the flower and fruit. If we examine the growth phase more closely, we see that the plant, as it vegetates and progresses from node to node, from leaf to leaf, is likewise carrying on a type of reproduction, which differs from that occurring in fruit and flower in that it is *successive* instead of sudden, appearing in a series of individual developments. (Goethe 1989, p. 76)

Formal unity underpins a dynamic of contraction and expansion. As Goethe continues, the distinction he draws above between vegetative growth and reproduction is a difference between a movement of *expansion* into stalk or stem and a movement of *contraction* of the parts of the plant into the concentrated form of the flower (Goethe 1989, p. 77). At this point, it seems to be a somewhat arbitrary choice to prioritize a reading here of *form* over the *processes* of metamorphosis, but I will return to this later.[5]

If we return to the problem of recapitulation (that ontogeny recapitulates phylogeny) the methods of analysis of distinct stages of development from the mammalian embryo to adulthood, like the identification of distinct organs and stages of plant development (Goethe), both presuppose the reducibility of the essence of the organism to fixed states or stages. Goethe, for example, documents each parts of the plant as some permutation, to different degrees of contraction and expansion, of an original or 'Protean' leaf form. One might use observational drawing, as Goethe and then Haeckel advocated,[6] or with more modern techniques in photography and time-lapse photography, to try to capture these stages. But I would suggest that, by (i) looking at Goethe's subsequent writings on the spiral tendency, and (ii) Bergson's theory of memory and duration (in the next chapter), we might identify not only the limitations of the formalist method of thinking, but also the possibilities inherent in it for overcoming those limitations.

[5]Goethe is clearly a forerunner of later debates in animal taxonomy. Another more famous formalist, and recognised figure in the history of biology, is Etienne Geoffroy Saint-Hilaire, whose achievements included the identification of the unity across species anchored in the vertebrate skeleton.

[6]See Robert J. Richards in Kort and Hollein (2009) who notes that 'Haeckel intended to represent not only the essential geometrical structures of radiolaria but their beauty as well, which he was able to portray through the use of color and the balanced arrangement of creatures in his atlas plates' (Kort and Hollein 2009, p. 99). Richards proceeds to attribute the influence of both Goethe and Alexander von Humboldt upon Haeckel's endeavour to recreate the naturalist's first encounter with the 'extraordinary beauty' of a living creature. The liveliness of the depiction must, in some way, recreate that of the original sight.

2 Spiral and Serpentine Lines

What are the 'spirals' that Goethe speaks of in his Botanical Writings?[7] Some clues can be found in Goethe's appeal to the term *homoeomeriae* from the fragments of the Presocratic thinker, Anaxagoras, whose view of nature included the affirmation that there is no empty space, and that the creation from nothing and destruction into nothing are impossible. He complains of his predecessors (such as Parmenides and Zeno) that they 'do not have a correct view of generation and destruction; for no thing is generated or destroyed; rather, they are mingled and dissociated from existing things. And for this reason they would be correct to call generation mingling and destruction dissociation' (Barnes 2001, p. 195). Thus, all material things are always made up of many other things, and when they appear to either come into being from nothing, or to be destroyed into nothing, it is only because we cannot physically see the smallest parts of which they are composed. The materials from which all things are made can therefore be said to be homoeomerous; they are 'uniformly what they are *through and through* and *all the way down*' (Sisko 2010, p. 445). Goethe uses this idea of homoeomereity in the formal biological description of spirals to refer to a part/whole relation that is indicated in, for example, the way in which stem cuttings can be taken from plants, replanted, and then grown as new individuals, thus revealing how '...independent life is attributed to them, also the power to move independently and to assume a definite direction' (Goethe 1989, pp. 128–129). This is also referred to as 'vital incurvation.'

Here, we are in fact presented with two interrelated tendencies, the vertical and the spiral. The vertical tendency describes the process from germination, taking root and building up stalks and fibres, and branches (Goethe 1989, p. 129). Crucially, we start to see an indication of Goethe's aim to overcome a purely physical or physiological description in favour of some sort of impetus, or a true tendency in process terms, where 'the vertical tendency should be looked upon symbolically as a

---

7This section refers to 'The Spiral Tendency' (1831), published one year before Goethe's death; together with a revised version of the 'Metamorphosis' essay (originally published in 1790); and 'On the Spiral Tendency in Plants' (1833; posthumously published from notes made in 1831).

staff, basic to existence and preserving it for a long period' (Goethe 1989, p. 131). The prolongation forwards, or upwards, is exemplified not just in rigid structures of stalks but also in the climbing or creeping forms of vines (e.g. in ivy and bindweed). In contrast to this, the spiral tendency refers to:

> the developmental, reproductory, and nourishing element. As such, it is temporary and almost independent of the vertical; operating in excess, it is soon exposed to ruin, and perishes; joining the vertical, it fuses with it to form a lasting union as wood or some other solid. (Goethe 1989, p. 132)[8]

The spiral tendency that seemed to preoccupy Goethe in the final years of his life was a recurrent image or theme in eighteenth century writings on aesthetics, starting perhaps with William Hogarth's essay 'The Analysis of Beauty' (1753). As Sabine Mainberger observes, Hogarth affirmed 'the existence of a standard smoothly curved line that can be found everywhere' across works of art of the Classical period, in leaves, flowers, and the bodies of animals in their basic forms and in their posture and movement (Mainberger 2010, pp. 203–204). Compare now Bergson's essay on 'The Life and Work of Ravaisson,' in which he cites Félix Ravaisson's particular appreciation of the drawings and underlying methodology of Leonardo da Vinci:

> There is, in Leonardo's *Treatise on Painting*, a page that Ravaisson loved to quote. It is the one where the author says that the living being is characterised by the undulous or serpentine line, that each being has its own way of undulating, and that the object of art is to render this undulation

---

[8]The sense of the spiral tendency that I am exploring has a rather different emphasis than Goethe's own, which associates the vertical and spiral with male and female characteristics: '...the vertical and spiral systems are closely bound together side by side in the living plant. When we see that the vertical system is definitely male and the spiral definitely female, we will be able to conceive of all vegetation as androgynous from the root up. In the course of the transformations of growth the two systems are separated, in obvious contrast to one another, and take opposing courses, to be reunited on a higher level' (Goethe 1989, p. 145). As ludicrous as the comparison sounds, even here the potential to overcome the gender stereotyping at play is contained in the fact of plant androgyny or hermaphroditism; ultimately, neither sex is dominant nor sufficient in isolation, but instead represents a duality abstracted from the androgynous whole.

> distinctive. [...] True art aims at portraying the individuality of the model and to that end it will seek behind the lines one sees the movement the eye does not see, behind the movement itself something even more secret, the original intention, the fundamental aspiration of the person: a simple thought equivalent to all the indefinite richness of form and colour. (Bergson 1992, pp. 229–230)

In the next section, I do not simply want to suggest that 'art' is the solution to the problem of presenting the priority of process over form (time over space). After all, Bergson's comments on intuition constitute his contribution to the debate about the role of philosophy in the wake of the rapid development and cultural ascendancy of the empirical sciences, a debate to which Kant's response was to establish the transcendental grounds of experience and thus the limits of philosophy. I do grant that there is merit in pursuing the visual significance of the spiral and serpentine forms, but I am interested here in the attempt to describe biological 'Process,' and the challenges this poses to written and diagrammatic modes of presentation. In this respect, Bergson's diagrammatic presentations of memory enact some of the problems I have summarized in Goethe (and the ideological lineage to which Goethe's botanical writings belong). Two aspects of Bergson's philosophy are brought to mind: (i) the critique of the cinematographical mechanism of thought, and (ii) the theory of intuition as the possibility of encountering duration freed from the fetters of the intellect.

i. Bergson's critique of the cinematographical illusion, as he names it in Chapter 4 of *Creative Evolution*, commences with a critique of the reductionism characteristic of formalist models of development, whereas 'in reality the body is changing form at every moment; or rather, there is no form, since form is immobile and the reality is movement' (Bergson 1998, p. 319). It is not any single one of the particular forms, but rather the process of continual change, that is real. In the perception of changes in the organic body we cannot help but take snapshot views on the process. Such changes tend to be so subtle that what emerges from our perception of them are 'a single mean image' that serves as a generalization of our real experience of continual, barely perceptible, change. This generalizing tendency is the source of essentialism. What Bergson calls here the cinematographical illusion is that aspect of human thought that enables the retrograde movement of truth, or the confusion of the method of analysis for the movement analysed. In his critique of this function, Bergson dismantles what for us, over one hundred years later, is now such a familiar process that perhaps the illusory quality of motion captured on film, and the layering of traditional celluloid with digital technology, is even more deeply embedded in our everyday understanding of the nature of perception than it was for Bergson's readers. Whilst the interventions of digital technology into our understanding of perception are interesting in themselves (and I explore this further in the next chapter), the analogy between the cinematograph (the camera) and the human intellect is still effective in making a very simple point about the secondary status we have tended to grant to process. For example, a film of a horse running across the screen, in its composition of a series of still photographs of it, gives us a sense of motion that is not contained in any of the successive images themselves, but in the *apparatus* that lends its own motion (its mechanical movement) to the static images. Consider Georges Canguilhem's subsequent critique of the hierarchy of machines over organisms (Canguilhem 2008). We cannot compare organisms to machines except somewhat tautologically. Machines that appear to be self-driving simply hide their dependency on some organic source of energy because they rely upon some form of energy-storage. The automaton relies upon this illusory status of the machine, and to compare animals to automata would simply be to compare them to other kinds of animals. The comparison defers the question of the *source* of motion. Bergson's analogy of the cinematograph anticipates Canguilhem's critique in the mistaken ascription of motion to a mechanism in which the real source of that motion is obscured. Likewise, and as a fundamental part of our processes of perception, our mind fixes movements into static images and then reconstructs them into a semblance of mobility: 'Whether we would think becoming, or express it, or even perceive it, we hardly do anything else than set going a kind of cinematograph inside us' (Bergson 1998, p. 323). Whilst the cinematographical mechanism of thought is an illusion, it is a *useful* illusion, according to which we carve up our surrounding reality in order to better navigate it. It does, however, mean that we risk reducing reality to a set of rigid concepts.[9]

ii. Bergson's theory of intuition opens up questions about our available strategies for overcoming the illusory mechanisms of thought. Bergson's perspective on the role of philosophy does not supplant, but rather complements, the function of the intellect. His writings on what this intuitive function might be are somewhat scattered and do not form a single, cohesive prescription for thinking about process. As Isabelle Stengers writes rather acerbically, there is a tendency amongst philosophers to place themselves at the terminal point of the traditions to which they respond. Bergson is guilty, on this view, of setting himself up as the unification of the tendencies of the intellect and instinct by trying to exemplify his own theory of intuition (Stengers 2011, p. 115). Yet Stengers' defence of Whitehead from this tendency should equally apply to Bergson's thought: what is important is not the status of any particular philosophy within the context of its history. What is interesting, rather, is an 'adventure of reason' that is 'not a definition of reason, but a particular description of the historical process that exhibits what we call "reason" as one of its questions' (Stengers 2011, p. 115). In Bergson's case, the historical process is an evolutionary one, in which life itself has evolved as an exploration of its own encounters with the world, manifested in a plurality of forms of life that each blend different degrees of intellectual and instinctive contact with things. As Bergson asks:

> What, then, is the principle that has only to let go its tension,—may we say to detend,—in order to extend, the interruption of the cause here being equivalent to a reversal of the effect? For want of a better word we have called it consciousness. [...] In order that our

[9]This is not only a matter of making category mistakes within the confines of philosophical or scientific reflection. It extends to our treatment of other species, in which such reductionism manifests, at its worst, in the forms of violence we visit upon them, reducing them to objects of study in the laboratory, or to units of production within an agricultural system. What possibilities might arise in our capacity to think, and in our attitudes towards other species, if we were to confront the tendencies of the intellect with the processual source of its concepts?

> consciousness shall coincide with something of its principle, it must detach itself from the already-made and attach itself to the being-made. It needs that, turning back on itself and twisting on itself, the faculty of seeing should be made to be one with the act of willing,—a painful effort which we can make suddenly, doing violence to our nature, but cannot sustain more than a few moments. (Bergson 1998, pp. 237–238)

An analysis of life, and of consciousness, is simultaneously an analysis of methods. In the final section of this chapter, I want to look at a number of potential criticisms of a primarily formalist approach in the methods of drawing and visualization from the perspectives of the sciences and the arts. Initially, we might simply assume that drawing, dealing in static images, allows us to reinforce the 'retrograde movement of truth.' The emphasis on form (the leaf, the vertebrate skeleton) encourages us to overlook the processes of development. Our descriptions are inadequate to growth whenever they settle on any particular state and declare that state either the origin or the *telos* of that organism's development. Are there possibilities contained in drawing, time-lapse photography, and animation that complicate the assessments of their hidden mechanisms (as both Bergson and Canguilhem identify)? Can we focus instead on what we might call a phenomenological relation or, as Whitehead would put it, a prehensive relation, between the artist and her work?

3 Drawing Lived Time

Scottish biologist and mathematician, D'Arcy Wentworth Thompson, in his book *On Growth and Form* (1917) explores a number of the problems broached so far, namely, the articulation of the interactions between internal and external selective forces, through a consideration of the relation between microscopic and macroscopic processes, and the use of explanatory principles between the extremes of teleology and physical reductionism. This final section will also serve as an outlook onto the next chapter on the potential of speculative methods in drawing and diagram-making as a way of exploring the methodological links between physics and metaphysics.

Thompson advocates morphology, the close study of living forms, grounded in Goethe's botanical studies, as a way of demonstrating the strict conformity of processes of growth and development in living things to physical laws, starting from their rudimentary formation in embryogenesis. The interactions of form and growth, being mathematical problems in the former case, and physical problems in the latter, mean that 'the morphologist is, *ipso facto*, a student of physical science' (Thompson 2016, p. 8). The book *On Growth and Form* deploys extensive examples and diagrams to demonstrate this fundamental feature of life, but it is also an invaluable meditation on the methods of the morphologist, and of the utility of visualization for working out conceptual problems more broadly. As with the formalism of Goethe, the priority given to form by no means ignores or obscures the dynamics of growth. We have already seen through Goethe's attempts to articulate the processes of metamorphosis and the spiral tendency that the identification of some sort of 'Protean' form necessitates a subtle account of the ways in which open-ended processes or tendencies of growth are 'reigned in' and shaped by internal and external physical determinants. The task of the morphologist is not to 'freeze' processes in their tracks, and thereby succumb to Bergson's cinematographical illusion, but rather to study the 'action of forces' that have produced changes in form over the course of an organism's life (or indeed of any non-living object). As Thompson underlines:

> The form of an object is a 'diagram of forces', in this sense, at least, that from it we can judge of or deduce the forces that are acting or have acted upon it: in this strict and particular sense, it is a diagram, –in the case of a solid, of the forces that *have* been impressed upon it when its conformation was produced, together with those that enable it to retain its conformation; in the case of a liquid (or of a gas) of the forces that are for the moment acting on it to restrain or balance its own inherent mobility. (Thompson 2016, pp. 11–12)

Expanding this formal principle to a living organism then incorporates both a *kinetics* of its physiological development and degradation and a *statics* or 'balance of forces' that allows it a certain degree of equilibrium (or relative permanence). The search for the *effects* of forces, in the absence of direct evidence of the nature of those forces (that, historically, might have been attributed to some Prime Mover or to a vital or animating principle), echoes Bergson's references to an *élan vital* in *Creative Evolution*, insofar as it is an epistemological and methodological claim about the limitations of conceptual knowledge. Thompson's criticism of 'the common language of books' within the realms of embryology and evolutionary theory that focuses excessively, in his view, on the 'material elements' that trigger change overlooks the effects of the *dynamics* of development (Thompson, p. 14). Predating the discovery of the structure of DNA, Thompson's perspective promotes a study of the interplay of forces and its effects on individual parts of an organism before the biological sciences' increased preoccupation with the detail of the genetic composition of an organism:

> Matter of itself has no power to do, to make, or to become: it is in energy that all these potentialities reside, energy invisibly associated with the material system, and in interaction with the energies of the surrounding universe. (Thompson 2016, p. 160)

Later complemented by Schrödinger's explanation of the connection between atomic and molecular phenomena, Thompson examines the differing effects of scale on organic development. The difference between the behaviour of a ripple in a body of water and that of a large wave illustrates this contrast in effects between molecular and molar forces, an observation of effects that can then be applied to living organisms. Here, smaller forms tend to arise in conformity with surface-tension, hence the recurrence of spherical forms in micro-organisms, whilst larger forms are more subject to the effects of gravity (Thompson 2016, p. 33).

What, then, is the relationship between form and growth? What are these molecular or microscopic forces, and how can we discern them mathematically and diagrammatically, according to Thompson? As he explains, any magnitude by itself can be represented by a number or by a line of definite length. Changes in magnitude (in space only or time only) can be represented by a line of varying length. This is called a vector. Changes in magnitude, if rendered in spatial terms, can be represented by two lines, for example, one denoting length, and the other breadth. The resulting picture or drawing of the object is, then, a 'plane projection' of that object. We can thus define form as a 'ratio of magnitudes' (Thompson 2016, p. 50). Furthermore, when variations in magnitude are mapped against successive moments in time, then this represents the object's 'growth,' and when we plot length (representing magnitude) against time, the resulting two-dimensional diagram gives us a 'curve of growth,' that is, it indicates the *velocity* or rate of growth (Thompson 2016, p. 51). The accumulation of such data then helps to build up a more sophisticated visualization, albeit an abstract one, of the development of the chosen object, such as a developing embryo. Specifically, the superimposition of 'plane projections' of the developing organism over time yields 'a three-dimensional diagram which is a partial representation (limited to two dimensions of *space*) of the organism's gradual change of form or course of development' (Thompson 2016, p. 51). Thompson argues that this intimate correlation between form and growth demonstrates that the study of organic development, understood in terms of rates of growth, or velocities, ought to be a prerequisite for the study of form (Thompson 2016, p. 51).

The physical determinants of the changes in form differ according to the relative size and complexity of the organism, a problem that Thompson answers in a criticism of a passage in Darwin's *Origin of Species*. Darwin's difficulty in mapping the potential dynamics of use and disuse onto variations in species of birds with a common ancestor lies in the phenomenon of the great resemblance between the foetuses of each species, whereas the forms of the adults differ greatly. Thompson counters Darwin's suggestion that selection (determined by use or disuse) tends to be enacted upon the adult members of a species. Instead, he reiterates the difference between physical effects on smaller forms and those that occur on the greater magnitudes involved when dealing with individuals in later stages of life. In short, the developmental changes in an embryo are more greatly determined by general molecular forces, whilst the changes in later stages are increasingly idiosyncratic as the individual grows in size and form. Such an insight, Thompson notes, refers us back to the phenomenon of recapitulation, favoured, as we saw, by Haeckel. If ontogeny recapitulates phylogeny, it is because early developmental processes are shaped within common physical constraints.

How are the actions of 'molecular forces' demonstrated, such that they help to explain what might otherwise seem to be the emergence of highly improbable structures? Thompson cites a number of examples and cases that utilize analogues of life in an effort to minimize the explanatory burden of the biologist, such as the comparison of 'caryokinesis' (the division of a cell's nucleus in the process of mitosis) with a bipolar electrical field. A simpler, and immediate visual analogue of this aspect of mitosis is also found in diffusion experiments using ink and salt water, so configured as to mimic to a remarkable degree the coalescence of chromosomes between the two parts of the just-divided nucleus. Such demonstrations reinforce the insight that strictly non-living physical forces are at play in the most fundamental of living processes. As Thompson continues, the cell is best described as 'a "sphere of action" of more or less localised forces; and of these, surface-tension is especially responsible for giving to the cell its outline and its morphological individuality' (Thompson 2016, p. 197). No particular part of the cell is identifiable as living as opposed to its non-living parts, since it is only in the interaction of the parts that energy-transfer takes place. Accordingly, the differentiating terms 'nucleus' and 'cell' should only be used 'in topographical expressions denoting two differentiated areas in a common structural basis' (Thompson 2016, p. 198). This characterization of the cell presents us not with a material continuity but rather 'a continuity of forces, a comprehensive field of force, which runs through and through the entire organism and is by no means restricted in its passage to a protoplasmic continuum' (Thompson 2016, p. 200).

We saw how Darwin, in *The Origin of Species*, marvelled at the mathematical precision displayed in the honey bee's construction of complex honeycombs (Darwin 1996, p. 183). Thompson explains that it is not an innate craft of the bee, but a much more rudimentary process involving the interaction of the bee's instinctive work with the physical principles of surface-tension and the tendency towards equilibrium:

> …the walls assume their configuration when in a semi-fluid state, while the papery pulp is still liquid, or while the wax is warm under the high

> temperature of the crowded hive. Under these circumstances, the direct efforts of the wasp or bee may be supposed to be limited to the making of a tubular cell, as thin as the nature of the material permits, and packing these little cells as close as possible together. It is then easily conceivable that the symmetrical tensions of the adjacent films (though somewhat retarded by viscosity) should suffice to bring the whole system into equilibrium, that is to say into the precise configuration which the comb actually presents. (Thompson 2016, p. 333)

The deceptively complex construction of 'Maraldi pyramids' by the bee is attributable to the same forces that result in the facets of a rhombic dodecahedron to be found in a collection of soap bubbles or in the complex of uniform cells in plant stems and leaves. The structures in the former case can be explained in the same way as those in the latter.

Thus, the particular configurations of living cells tending towards spherical form, the complex processes of mitotic division, and the recurrence of certain geometrical forms across all orders of life are explicable in terms of a purely physical consideration of the interactions of form and growth. As we saw in the last chapter, Prigogine's later experiments in the production of dissipative structures would carry forward this intuition about emergent properties. What Thompson's analysis achieves is a potential simplification of certain problems within evolutionary and developmental biology. If the early stages of embryonic development can be explained with reference to basic physical principles, then this simplicity can minimize the need for overly speculative explanations for the emergence of more complex structures in adult forms and the ecological relationships between different species. Thompson's recourse to morphological study, mathematical models, and diagrammatizing demonstrates an intricate relation between the processes of observation, imaging, and problem-solving, and yet as contemporary British artist Gemma Anderson notes, 'morphological analysis as a taxonomic tool has been eclipsed by DNA and genomic analysis' resulting in a decline in the practice of the observational drawing of specimens (Anderson 2014, p. 233).

Anderson's work in collaboration with a number of scientists (as well as, most recently, with philosopher of biology, John Dupré), has explored the value of morphological drawing as a way of addressing this increasingly under-utilized practice within zoological taxonomy. The practice of morphology should be seen, Anderson claims, as an invaluable tool both pedagogically and professionally for scientists and artists alike (Anderson 2014, p. 233). As we saw in the previous chapter, the increasing focus on DNA analysis, using ever more sophisticated tools such as the electron scanning microscope, as well as the increased focus on statistical models of biological functioning, attesting to a sharpening of a broadening range of scientific specialisms, have been noted by Mayr, Margulis, and Lovelock alike, as detrimental to the depth of scientific discovery. Scientists working on different sub-disciplines are becoming less able to communicate and compare results with one another, whilst Anderson speaks of a much more local phenomenon of alienation experienced by the individual whose taxonomical work relies upon DNA sequencing, in place of observation of the specimen in question:

> it is not uncommon for molecular biologists to consider only the molecular data without any other knowledge of the species. Quentin D. Wheeler identifies this disconnection as a problem: 'Much data may be collected but what is lost is the greater understanding of what those data mean or don't mean.' (Anderson, 2014, p. 237)

Anderson's initial study of the value of drawing, 'an intimate, devotional act of wonder at the many forms and puzzles that species present' (Anderson 2014, p. 239), points to the power that it has, *and the time that it takes*, to provoke new realizations about the specimen being studied. There is a quality that is unique to the *act* of drawing that enables thought to articulate itself along lines not accessible through conventional signs (be they geometrical, diagrammatical, or linguistic). The development of the potential of this view of 'drawing as a way of thinking' in what Anderson coins 'Isomorphology' points to some potentially intriguing lines of research for the philosopher as well as for the artist and scientist. Whilst Anderson's most recent collaborations have involved the work of John Dupré, the focus has been on the interactions between the practices of scientists and artists, with, it seems, the theoretical scaffolding provided by the philosophy of science. In line with the broader explorations of the present book, I am interested in Anderson's work for its implications for how we work in philosophy itself. A first step to thinking about this is the project of Isomorphology which is 'the study of the shared forms and symmetries of animal, mineral and vegetable species through drawing practice' (Anderson 2016, p. 10). The practice incorporates a kind of 'classificatory pluralism' (a term Anderson borrows from Dupré) that liberates taxonomy from adherence to fixed lines of demarcation between species, and indeed kingdoms, and ultimately non-living physical kinds. Anderson stresses the potential that such an approach might have for provoking new lines of enquiry in the work of classification, unfettered as it is by conventional specific and formal boundaries. Crucially, 'Isomorphology' is

> a blending of scientific and artistic experimentation which brings with it new modes of seeing and classifying the natural world. By placing the making of observational drawings at the foundation of this artistic experimentation, Isomorphology demonstrates drawing's continued viability as an epistemic process and as a way of producing knowledge. (Anderson 2016, p. 11)

The development of Anderson's work in collaboration with both mathematicians and biologists offers some conclusions about the potential applications of an approach centred on the act of drawing to philosophical enquiry itself. A clue to the approach we might take can be found in Anderson's work with mathematicians to develop a 'visual vocabulary' for presenting imaginary objects that exist only in the combinations between geometrical formulae and the minds' eyes of the mathematicians attempting to articulate them. Whilst the approach to theorizing the relationship between the practices of thinking and drawing is manifestly dualistic, being founded on the activities of the 'The Thinker' and 'The Drawer,' the resulting artworks speak to a more continuous and evolving relationship between the two conventionally distinct practices of mathematics and art:

> we focus on the drawing of imaginary objects— that is, objects that we see with our mind's eye. Whether the drawn object be physical or imaginary, all drawing is a sort of inverse vision. By drawing with pencil on paper we give physical form to our mental images, and in the process we learn to see them better. Thus, in this context, drawing is a tool to train ourselves to see imaginary things better. (Anderson et al. 2015, p. 440)[10]

Anderson reiterates the value of the act of drawing as a pedagogical and creative process, rather than as a by-product of work that is to be ultimately translated back into abstract formulae:

> In each case, the process of drawing and the reflection that accompanies it transforms the drawer, changing the way in which they know and understand the object that is being drawn. This explains why drawing is so much more effective in this context than the use of computerbased visualization software: It is the act and experience of drawing itself that creates intuitive understanding. (Anderson et al. 2015, pp. 441–442)

In her book, *Drawing as a Way of Knowing in Art and Science* (first published in 2017), Anderson reflects on the transformation of her own practice towards what she names 'Isomorphogenesis,' which is intended as an amalgamation of D'Arcy Thomson's 'grid transformations,' the use of colour gradation by Paul Klee, and FormSynth, the computer-modelling system developed by William Latham (Anderson 2019, p. 189). Anderson argues that Paul Klee's practice and artworks can be read as a kind of morphology, inspired by Goethe's methodology, and developing its own visualization of objects-in-process, in particular his use of gradations of colour (using layers of watercolour paint) to signify the passage of forms through a developmental sequence (Anderson 2019, p. 144). Combining some speculative work on the possibility of rendering three-dimensional objects in four or more dimensions with her readings of Goethe, Klee, and Thompson, Anderson's method of Isomorphogenesis proposes 'a kind of drawing algorithm involving drawing

[10]This is to be distinguished from 'the visual and spatial representation of nonvisual scientific objects, for example on the visualization of statistical data or on the visual representation of processes and of relationships between concepts' (Anderson et al. 2015, p. 440).

actions (verbs) performed on a set of primitive shapes rather than from observation. The algorithm simulates possible analogs of developmental series based on principles similar to those that regulate plant and animal growth' (Anderson 2019, p. 176).

The later application of this approach to the visualization of mitotic processes is, again, promoted on the basis of its value 'both in communicating the dynamic nature of biological processes and in generating new insights and hypotheses that can be tested by artists and scientists' (Anderson et al. 2019, p. 1). Anderson's work here in particular recalls D'Arcy Thompson's identification of forms as 'diagrams of forces' and develops the morphological attempt to articulate the interaction between form and growth in the developing organism. Anderson's drawings of the *process* of mitosis as a continuous set of changes in a single image, rather than a succession of images as snapshots of stages of the process, effectively combine the mathematical plotting of velocity, as explained by Thompson above, and the non-living chemical analogues of mitosis (the experiments in artificial caryokinesis) that mimic the appearance of mitotic division, as a means of reflecting on the physical dynamics that subtend the living process. The effort to visualize a process will never capture the true dynamic of the whole, but it can carry thought some way closer to an appreciation of its complexity. A widening of Anderson's approach to incorporate the working methods of philosophers would be an intriguing extension of this project.

Over the course of this chapter, I have looked at Goethe's altering emphases in his explorations of plant metamorphosis as offering a potentially non anthropomorphic vocabulary for discussing life in process terms. The idea of 'recapitulation' in the history of evolutionary and developmental biology is problematized in Goethe's botanical writings, precisely in the shift towards thinking in terms of tendencies, provided we see metamorphosis as a way to *overcome* formalism (and the androgyny of plants as a way to overcome his gendered descriptions of the vertical and spiral tendencies). The irreducibility of the microscopic to the form of the macroscopic (individual development does not necessarily echo phylogenetic development) problematizes the reduction or resolution of multiplicity into unity. In the next chapter, I will explore further the use of static diagrammatic representations of dynamic processes, and the extent to which they must always refer beyond themselves and, ultimately, break down. The complexity of particular conscious operations calls for the formulation of new concepts, or at least a resistance to monism. Thus, Goethe shows us both the potential problems and possibilities for thinking lived time through a non-anthropomorphic example of plant growth, the need to be cognizant of persisting Romantic ideals of unity of forms, and the formulation of complementary tendencies. Ultimately, we can view the form of the *protean leaf* as subordinate to the processes of metamorphosis, and the vertical and spiral tendencies. The spiral tendency is to be understood as less a visual form, than it is a remnant of an activity and a capacity for change. The challenge is to think about the physical form of the plant in terms of the primacy of process, be it what Goethe calls 'Metamorphosis' or 'Tendency.' The practice of drawing as a way of thinking (e.g. the possibility of an intuitive, unfettered drawing of four-dimensional 'visual objects') can be extended, for example, to a reading of Bergson's use of diagrams in his work: not as fixed representations of well-defined things, but as indications of processes that are continuously changing, whilst settling momentarily into fixed forms for the purpose of practical activity. They reveal *pure memory* to be a useful myth, so to speak, insofar as its suggested ontological status (as Deleuze, for example, emphasizes it) serves a particular function: to open up reflection on the reconfiguration of matter in temporal terms and thus the reconfiguration of mind in terms of *processes.* I will consider these questions concerning the visual presentation of the theoretical formulation of living or conscious processes in more detail in the next chapter.

References

Anderson, Gemma. 2014. Endangered: A Study of Morphological Drawing in Zoological Taxonomy. *Leonardo* 47 (53). https://doi.org/10.1162/leon_a_00675.

Anderson, Gemma. 2016. Drawing Resemblance: Mineral Nicknames and Isomorphology. *Architectural Theory Review*. https://doi.org/10.1080/13264826.2016.1177102.

Anderson, Gemma. 2019. *Drawing as a Way of Knowing in Art and Science*. Bristol: Intellect.

Anderson, Gemma, Dorothy Buck, Tom Coates, and Alessio Corti. 2015. Drawing in Mathematics: From Inverse Vision to the Liberation of Form. *Leonardo* 48 (5). https://doi.org/10.1162/leon_a_00909.

Anderson, Gemma, John Dupré, and James D. Wakefield. 2019. Drawing and the Dynamic Nature of Living Systems. *eLife* 8. https://doi.org/10.7554/elife.46962.

Barnes, Jonathan. 2001. *Early Greek Philosophy*. London: Penguin.

Bergson, Henri. 1992. *The Creative Mind: An Introduction to Metaphysics*, trans. Mabelle L. Andison. New York: Citadel Press.

Bergson, Henri. 1998. *Creative Evolution*, trans. Arthur Mitchell. New York: Dover.

Canguilhem, Georges. 2008. *Knowledge of Life*, trans. Stefanos Geroulanos and Daniela Ginsburg. New York: Fordham University Press.

Darwin, Charles. 1996. *The Origin of Species*. Oxford: Oxford University Press.

Deleuze, Gilles, and Félix Guattari. 1994. *What Is Philosophy?* trans. Graham Burchill and Hugh Tomlinson. London: Verso.

Esposito, Maurizio. 2014. Problematic 'Idiosyncrasies': Rediscovering the Historical Context of D'Arcy Wentworth Thompson's Science of Form. *Science in Context* 27 (1): 79–107. https://doi.org/10.1017/S0269889713000392.

Goethe, Johann Wolfgang. 1989. *Goethe's Botanical Writings*, trans. Bertha Mueller. Woodbridge, CT: Ox Bow Press.

Gould, Stephen Jay. 1977. *Ontogeny and Phylogeny*. Cambridge, MA: Harvard University Press.

Khandker, Wahida. 2014. *Philosophy, Animality and the Life Sciences*. Edinburgh: Edinburgh University Press.

Kort, Pamela, and Max Hollein, eds. 2009. *Darwin: Art and the Search for Origins*. Cologne: Weinand.

Lovejoy, Arthur O. 1964. *The Great Chain of Being: A Study of the History of an Idea*. Cambridge, MA: Harvard University Press.

Mainberger, Sabine. 2010. 'In the Vortex of the Spiral Tendency'—Questions of Aesthetics, Literature and Natural Sciences in the Work of Goethe. *Estudos Avançados* 24 (69): 203–218.

Sisko, J.E. 2010. Anaxagoras on Matter, Motion, and Multiple Worlds. *Philosophy Compass* 5: 443–454. https://doi.org/10.1111/j.1747-9991.2010.00313.x.

Stengers, Isabelle. 2011. *Thinking with Whitehead: A Free and Wild Creation of Concepts*, trans. Michael Chase. Cambridge, MA: Harvard University Press.

Thompson, D'Arcy Wentworth. 2016. *On Growth and Form.* Stellar Editions: USA.

5
Reminiscence

In the first chapter of *Matter and Memory*, Bergson describes the emergence, from an aggregate of images, of one privileged image that we might call a 'centre of real action' (Bergson 1991, p. 5). We are dealing here with a speculation on the nature of the material universe, and the terms with which we might describe its most rudimentary activities, on a scale of the most unreactive to the most active of bodies or 'images':

> Images themselves, they cannot create images; but they indicate at each moment, like a compass that is being moved about, the position of a certain given image, my body, in relation to the surrounding images. In the totality of representation they are very little; but they are of capital importance for that part of representation which I call my body, since they foreshadow at each successive moment its virtual acts. There is then only a difference of degree, there can be no difference in kind, between what is called the perceptive faculty of the brain and the reflex functions of the spinal cord. (Bergson 1991, p. 10)

On first reading of this text, Bergson's language is mystifying, though on later readings, and in the light of the works of other process philosophers,

W. Khandker, *Process Metaphysics and Mutative Life*,
Palgrave Perspectives on Process Philosophy,
https://doi.org/10.1007/978-3-030-43048-1_5

most notably Alfred North Whitehead, this image of images resonates within the history of ideas of a pre- or non-subjective world at the foundation of experience. The experience that is described in this account is thus one of an emerging and slowly-forming (and ever re-forming) subject. The centre of real action, accompanied by its zone of indetermination, resists its coalescence into a personal 'I' in order to better reflect the continual exchange of affect and influence between different 'images,' Bergson's term for matter. The experience being described here could be that of first awakening from sleep: surrounding images commence ablur, and then gradually the colours and shapes become invested with the significance, and the uses, they have for the awakening—soon to be acting—image they surround.

In the last chapter, I surveyed some of the uses of drawing as part of scientific research across works of a number of thinkers where Bergson lent a sceptical voice to the exploration of methods for the expression of process. I continue now with a more detailed analysis of the possibilities of 'drawing as a way of thinking,' across examples from Bergson's philosophy and contemporary physics. Specifically, as I indicated in the opening of the previous chapter, I commence with a comparison I have often used, for pedagogical purposes, in order to communicate the impersonal quality of Bergson's description of the 'privileged image': a comparison with the distorting effects of a black hole on its surroundings (and common visual renderings of this effect).[1] This may seem an unlikely analogue, but it has proved useful as a visual aid and complement to Bergson's own diagrams featured in *Matter and Memory*. The curvature or enfolding of spacetime effected by the singularity (at the heart of the black hole and hidden from view just beyond its horizon) at least superficially echoes the distortions of images surrounding the centre of action: the zone of indetermination. As a centre of action, I cannot help but distort, that is, interpret, my immediate surroundings in the light of my possible actions upon them. I can walk in this direction, but

---

[1] See Donald Marolf on the pedagogical use of embedding diagrams in physics: 'Fig. 1 shows how the equatorial plane of a black hole would be curved if, instead of lying in the black hole spacetime, it were part of familiar flat Euclidean three-space. Mathematically, this picture is said to represent an *embedding* of this plane into three-dimensional Euclidean space' (Marolf 1999, pp. 919–920).

not in the other (due to some obstacle), I can reach out for that handle and open the door, and so on. This is the interposition of memory on perception and action that serves as the core of Bergson's theory of consciousness. In many ways the analogy is, of course, problematic. It is, for a start, overly simplistic. The correspondence I suggest between the acting subject and the curvature of spacetime is limited to a relatively small degree of influence exerted by the 'centre' upon the physical appearance of the surrounding environment: for Bergson, in the universe as aggregate of images; for the physicist, in spacetime. Yet the comparison is, as I hope to show in this chapter, potentially much more complex and useful for thinking about our *methods* for understanding time and different kinds of processes. From a historical point of view, one cannot make the above connection without also acknowledging and examining the debate that took place between Bergson and Einstein in the 1920s on the metaphysical implications of the latter's theory of Special Relativity (SR), and outlined in Bergson's *Duration and Simultaneity* (1922). Historical reflections on the debate reflect unfavourably on Bergson for his 'infamous mathematical errors' (Scott 2006, p. 189), in contrast to the ease with which Einstein was able to dismiss Bergson's appeal to a 'philosopher's time'[2] (Olma 2007, p. 127). What we can see in Bergson's critique of SR, that on the face of it is a disagreement between theories of a single against a plurality of times, are its roots in his broader philosophical project to recover the reality of objectivity (in response to Kant), to reflect on the purpose and future of philosophy, and to think critically about, and with, the sciences in their endeavour to understand the physical universe.[3]

In this chapter, then, I commence with an overview of Bergson's response to the theory of Special Relativity in order to explore a number of problems, starting with the presentation of conscious activity in diagrammatic form and the difficulties that it raises. This follows from the

[2]However, the claim of Olma's paper is that attitudes towards 'qualitative time' amongst contemporary physicists may be changing.

[3]It is also worth noting that Bergson focuses only on SR rather than General Relativity (through which the existence of phenomena such as black holes are articulated), stating that he does not seek to refute Einstein's theories, but only to temper them with a number of metaphysically-driven problems.

previous chapter's explorations of visualization as integral methodological tools for understanding plant growth (in Goethe's case), through to Thompson's mathematically grounded diagrams, and Anderson's artistic speculations on methods and processes alike. In the present chapter, I want to respond to Bergson's criticisms of certain consequences of visualization, expressed in terms of the cinematographical illusion, through a study of his own use of diagrams in *Matter and Memory* and how these might be reconciled with more recent techniques for analysing and communicating contemporary research in relativistic physics. For this purpose, I refer to writings of the contemporary physicist Kip Thorne. The first parts of this chapter provide an overview of key problems leading up to Einstein's formulation of Special Relativity, followed by Bergson's reading of them, his critique, and the reasons for that critique found in both *Duration and Simultaneity* and other works in which Bergson sets up his methodological project. What Bergson sees as the role of metaphysics, in response to physical formulations of time, is to reassert the value of our real experiences of time as integral to an understanding of time or 'duration' itself.

The remainder of the chapter will be devoted to an exploration of the implications of a problem that Bergson sets up in *Matter and Memory* concerning the relation of cerebral matter to memory (or mind). Memories, he contends, are not stored in the brain. Deleuze's extension of this formulation, invoking a 'nonpsychological reality,' or a kind of being that grounds consciousness (or our psychological reality) expresses the ontological implications of Bergson's already contentious statement: it is not a reference to a mystical realm of the soul, but rather to the inadequacy of our concept of matter that, when applied to the problem of the mind's relation to the brain, leads to a reduction of living or conscious processes to the physical structures that we can observe. I suggest that a second look at Bergson's use of diagrams, on the one hand, and attempts in contemporary physics to visualize the structure and behaviour of otherwise 'imperceptible' objects such as black holes, on the other, provides us with a useful perspective on the purpose and limitations of diagrams and drawing as a way of thinking through process-philosophical problems. Whilst Bergson's *Duration and Simultaneity* focuses only on Special Relativity, the existence of things such as black holes (along with worm holes, or Einstein-Rosen bridges, and gravitational waves, etc.) are the consequences of the broader theory of General Relativity. I take advantage of this key difference to suggest some points of convergence and possibilities for thinking the philosophical and physical theories together, both in order to 'reconcile' the positions of Bergson and Einstein (if rather obliquely) and to solve some problems within certain philosophical interpretations of Bergson's theory of memory, in particular the virtual-actual relation and the coexistence of the past and the present visualized through the cone diagram. For examples of the utility of visual models in physics, I will make special reference to Kip Thorne's collaborations with visual effects technicians on the 2014 film, *Interstellar*.

1 Physical Time

When reflecting on the nature of time, using traditional concepts of movement and change, Bergson remarks that

> One might as well discourse on the subject of the cocoon from which the butterfly is to emerge, and claim that fluttering, changing, living butterfly finds its *raison d'etre* and fulfilment in the immutability of its shell. On the contrary, let us unfasten the cocoon, awaken the chrysalis; let us restore to movement its mobility, to change its fluidity, to time its duration. (Bergson 1992, p. 17)

Faced with a theory that would precipitate a revolution in the physical understanding of time and motion, Bergson, the renowned philosopher of time, would surely be well placed to comment on the rendering of time as relative and plural by Albert Einstein's formulation of Special Relativity, but it remains a controversial intervention. Bergson's analysis of Special Relativity starts with the experiments of Albert Michelson (in 1881; later versions were to be conducted together with Edward Morley) regarding the propagation of light. The relation between the keystones of Newtonian physics, absolute space, absolute time, and an 'aether,' were tested in these experiments. Specifically, the effect of the postulated aether on light travel could be tested by measuring variations in light speed from the point of view of a stationary observer. If space and time are each absolute, and light travels at the same speed in all directions, then under certain conditions, some variation in the perceived speed of light should be detectable. A stationary observer watching light travel should see it propagate or transfer at the same speed in all directions. Physicist Kip Thorne helpfully evokes the analogous effects of travelling at speed from the perspective of a passenger on a moving train. The passenger watching birds flying in the same direction of the train will perceive the birds as flying more slowly, whilst birds flying in the opposite direction to the moving train will appear to speed up. Generalized outwards to think about the relation between celestial bodies in motion (the moving train as the Earth in orbit around the Sun), an equivalent effect on our perception of light travel (the birds in flight) should occur from our 'stationary' standpoint as observers on Earth. The perceived speeding up or slowing down of light should occur because light, according to Newton's laws, has an absolute propagation speed regulated by the 'aether':

> Beating its electric and magnetic fields against the aether, light propagates always at the same universal speed through the aether, regardless of its propagation direction. And since the aether (according to Newtonian concepts) is at rest in absolute space, anyone at rest will measure the same light speed in all directions, while anyone in motion will measure different light speeds. (Thorne 1994, p. 63)

The assumptions of Newtonian physics above, in which a moving object must experience a slowing or speeding up of light travel, imply that any change in the experience of the speed must be measurable from the moving position of the Earth, no matter how slow it is in comparison to the speed of light; in other words 'we on Earth should measure the speed of light to be different in different directions, and the differences should change with the seasons—though only very slightly (about 1 part in 10,000)' (Thorne 1994, p. 64). However, the Michelson-Morley experiments detected no such differences. The ensuing attempts to explain these repeated null results included the insight, proposed concurrently by Hendrik A. Lorentz and George F. Fitzgerald, that bodies moving in space must contract very slightly in length.[4] This would challenge the long-held view that material bodies retain their shape and size in relation to absolute space, regardless of their speed of motion. Einstein's response (1905) to the results of the experiment were even more radical, doing away with the Newtonian precepts of absolute space, absolute time, and the aether altogether. A statement of the new foundations we can now refer to as Special Relativity (SR) retains only the absolute speed of light. There can be no privileged observer; all positions in space and time are relative to one another (that is, space and time are relative insofar as they are mutually generated by co-existing bodies in motion) (Thorne 1994, pp. 72–73). We will see that this eradication of a privileged observer is too swift a move from the perspective of Bergson's analysis.

Contra Einstein's claim of the interchangeable nature of positions in space, Bergson objects that the collectively generated solution to the Michelson-Morley experiments (the upending of Newtonian absolute space and time), states something self-contradictory about the status of the physicist as observer. The cancelling out of the null result of the Michelson-Morley experiments, compensated by contracted matter and expanded (dilated) time, discounts the experience of time and movement of the *moving* observer:

> In short, considered in space and time, system *S'* is a double of system *S* which, spatially, has contracted in the direction of its motion, and, temporally, expanded each of its seconds; and which, finally, has broken up into succession in time every simultaneity between two events whose distance apart has narrowed in space. But these changes escape the observer who is part of the moving system. Only the stationary observer is aware of them. (Bergson 1999, p. 11)

Bergson will go on to agree with all of the steps taken to arrive at the three basic principles of SR, but with a modification to the status

[4]As Thorne explains, only 'a tiny (five parts in a billion) contraction of length along the direction of the Earth's motion could, indeed, account for the null result of the Michelson-Morley experiment' (Thorne 1994, p. 65). See also Lorentz (1952), p. 5: 'one would have to imagine that the motion of a solid body (such as a brass rod or the stone disc employed in later experiments) through the resting ether exerts upon the dimensions of that body an influence which varies according to the orientation of the body with respect to the direction of motion.'

of the plural times added to the time of the stationary observer. Thus despite the proclamations of Relativity Theory that there is no privileged observer, it quietly privileges the position of the physicist at any chosen point whatever in spacetime. In this unacknowledged privileging, or perhaps 'unconscious metaphysics', of the observer Bergson exposes the way in which 'a single time and an extension independent of duration continue to exist in Einstein's theory considered in its pure state; they remain what they have always been for common sense' (Bergson 1999, p. 19). Furthermore:

> [if] we subscribe to Einstein's theory, the multiple times will remain; but there will never be more than a single real one among them, as we propose to demonstrate; the others will be mathematical fictions. That is why, in our opinion, if we adhere strictly to Einstein's theory, all the philosophical difficulties relative to time disappear, and so too will all the oddities that have led so many minds astray. (Bergson 1999, p. 20)

It would be easy for the present-day commentator on this history to concede (presumably on Bergson's behalf!) Einstein's 'victory' over Bergson, and this commentator certainly does not wish to simply repeat or defend any modification of Einstein's insights or detract from his continued significance in contemporary physics. Indeed, as the rest of this chapter will show, I am interested in 'thinking with' the later extension of SR into General Relativity with its speculations on spacetime curvature. However, what I do want to stress is the importance that Bergson's objections to the proliferation of times have in relation to his broader philosophical project. At stake in the proposed explanation of the Michelson-Morley experiments is a problem that lies at the very heart of Bergson's theory of duration: the fact that each of us, insofar as we wish to engage in metaphysics, has immediate access to the 'data of consciousness.' We do not need to defer to the evidence of experiments conducted by anyone else in order to inspect for ourselves the manner in which memory, perception, and action proceed. On this point, Bergson refers to a correspondence between Descartes and More on the nature of reciprocal motion (Bergson 1999, p. 21). For Descartes, two moving objects can be described as displaced in relation to one another (we can adopt the perspective of either object, and regardless of perspective, the other will seem to be moving). For More, the 'bad physicist' (Koyré 1968, p. 142), on the other hand, in certain circumstances, it is quite clear that the motion (signalled through effort) belongs to one of the bodies: the man out of breath, with reddened face.[5] What this mildly amusing dispute reveals is a serious point about the role of the metaphysician, who must 'penetrate into the interior of things; and the true essence, the underlying reality of a motion can never be better revealed to him than when he performs the motion himself, when he doubtless still perceives it from the outside like any other motion, but in addition apprehends it from within as an effort whose track alone was visible' (Bergson 1999, p. 22).

2 Metaphysical Time

In his two-part Introduction to *The Creative Mind*, Bergson recalls the trajectory of his own philosophical education, framed within the problem of the 'retrograde movement of the true growth of truth'. This movement of the truth, and its connection to the very purpose of intelligent life, is vital for an understanding of the stance Bergson adopts in relation to the implications of SR. How time or Duration has been understood in the history of ideas (across philosophy and science) stands as an exception to other objects of measurement. It is a special case because time itself is not the thing measured, but a trajectory or a line that remains only in the wake of a time that has already passed: 'what is counted is only a certain number of extremities of intervals, or *moments*, in short,

[5]See Alexandre Koyré (1968) for an assessment of different elements of this debate: 'In other terms, relative motion implies absolute motion and can only be understood on the basis of absolute motion and thus of absolute space. Indeed, when a cylindrical body is in circular motion, all its internal points not only change their position in respect to its surrounding surface, or a body placed outside it: they move, that is, pass *through* some extension, describe a trajectory *in* this extension which, therefore, does not move. Bodies do not take their place with them, they go from one place to another' (Koyré 1968, pp. 144–145). This objection will be echoed, to some extent, in Bergson's criticisms of Einstein, with the modification that it is an absolute or single *time* that is presupposed in the theory of a relative and plural time.

virtual halts in time' (Bergson 1992, p. 12). Bergson accepts that the succession of moments organized on a line are the legitimate objects of study for the mathematician, the physicist, and the astronomer, for whom the condensations and dilations of time can be imagined freely. However, as More's response to Descartes illustrates, for a living conscious individual a time dilated would be a source of fatigue, a condensed time perhaps of relief (Bergson 1992, p. 13). What, Bergson asks, is this lived duration?

> Suppose we try to find out what it is?—How would it appear to a consciousness which desired only to see it without measuring it, which would grasp it without stopping it, which in short, would take itself as object, and which, spectator and actor alike, at once spontaneous and reflective, would bring ever closer together—to the point where they would coincide—the attention which is fixed, and time which passes? (Bergson 1992, p. 13)

However, this is not say that the effort of introspection is easy: far from it. Bergson's theory of the reciprocal genesis of intellect and matter underlines how intelligence feels most at home in 'inert matter' (the unorganized; as opposed to an attention to 'organized' vital processes); in the inspection of things external to itself that are easily matched to the immobility of its own concepts and language. In introspection, on the other hand, the mind feels as if it is in a 'strange land' (Bergson 1992, p. 41), going against its 'natural bent' (Bergson 1992, p. 41). The authority of the experiencing individual can only be a beginning. It is repeatedly invoked across Bergson's earlier texts, from the detailed studies of subjective experience in *Time and Free Will* (*An essay on the immediate data of consciousness*), to the invitations to the reader to problematize dualistic and psychological accounts of mind in *Matter and Memory* and *Mind-Energy*. For example, one could test the difference between strong and weak sensations, through an action such as applying pressure to one's arm, and gradually loosening that pressure to illustrate and to contemplate the claims of a theory that memories are simply weakened perceptions. In fact, through our concrete experience, as the experiment is intended to show, we can see that memory and perception must be different in kind[6] (Bergson 1975, p. 161). This recourse to the immediacy of experience as the foundation of the metaphysician's practice is also, of course, part of Bergson's response to Kant's critical philosophy, not as a refutation nor simply as an extension of it. Rather, Bergson tackles the problems identified by Kant himself in the first pages of the *Critique of Pure Reason.*[7]

If Bergson is to recover some metaphysical authority over the question of the nature of time, it must, after Kant, set out its bounds and its methodology. The aim, indeed the 'duty,' of philosophy according to Bergson is to:

> lay down the general conditions of the direct, immediate observation of oneself by oneself. This inner observation is warped by habits we have developed; the chief example of this warping is doubtless the one which created the problem of liberty—a pseudo-problem born of a confusion of duration with extension. (Bergson 1992, p. 27)

In the identification of the delimiting habits of thought, we can move beyond a Kantian resignation to our fundamental alienation from things in themselves and the accompanying speculation that only an intuitive faculty that eludes us could connect us to things. As Bergson contends, 'at least a part of reality, our person, can be grasped in its natural purity' (Bergson 1992, p. 28). To begin with, when we inspect the activity of our own consciousness, we see echoes of an activity that is not only shared

---

[6]Cf. also Bergson (1991), pp. 136–137, on the difference between strong and weak states, on the one hand, and perceptions and memories on the other.

[7]'It is a call to reason to undertake anew the most difficult of all its tasks, namely, that of self-knowledge, and to institute a tribunal which will assure to reason its lawful claims, and dismiss all groundless pretensions, not by despotic decrees, but in accordance with its own eternal and unalterable laws. This tribunal is no other than the *critique of pure reason*' (Kant 1929, pp. Axi–xii). And further: 'This attempt to alter the procedure which has hitherto prevailed in metaphysics, by completely revolutionising it in accordance with the example set by geometers and physicists, forms indeed the main purpose of this critique of pure speculative reason. It is a treatise on the method, not a system of the science itself. But at the same time it marks out the whole plan of the science, both as regards its limits and as regards its entire internal structure' (Kant 1929, pp. Bxxii–xxiii).

with other species, but one that characterizes the nature of our physical bodies. This activity is, perhaps surprisingly, that of abstraction or generalization. In retort to John Locke who affirmed that 'the faculties of brutes can by no means attain to abstraction' (Berkeley 1975, p. 79), George Berkeley observed that if this were the defining characteristic of animals, then 'a great many of those that pass for men must be reckoned into their number' (Berkeley 1975, p. 79). Bergson adds another dimension to this exchange when he observes the following:

> Every living being, perhaps every organ, every tissue of a living being generalizes, I mean classifies, since it knows how to gather, in the environment in which it lies, from the most widely differing substances or objects, the parts or elements which can satisfy this or that one of its needs; the rest it disregards. Therefore it isolates the characteristic which interests it, going straight to a common property; in other words, it classifies, and consequently abstracts and generalizes. (Bergson 1992, p. 54)

It is not, in fact, some eminently human faculty that organizes sensory information, but rather an embodied tendency to distinguish, say, sources of food (a meadow for a cow) or shelter (a stable for a horse). In human beings there may be some multiplication of alternatives by virtue of the accumulation of useful habits, but generalization is enacted at many levels: biologically (in organized matter), by generations of organic bodies in relation to environment and internal mutation (evolutionary process), by individual organisms; in physical properties (in unorganized or inert matter—sensory, elementary, force, etc.); and of course in human fabrications (Bergson 1992, p. 58). Furthermore, we can imagine this activity of condensation of any multitude of perceptions to make up any set of generalities we like. The greater the ability of that being to condense, the more of the universe it can encompass (Bergson 1992, p. 59). Take this example of the perception of colour in *Matter and Memory*:

> May we not conceive, for instance, that the irreducibility of two perceived colours is due mainly to the narrow duration into which are contracted the billions of vibrations which they execute in one of our moments? If we could stretch out this duration, that is to say, live it at a slower rhythm, should we not, as the rhythm slowed down, see these colours pale and

> lengthen into successive impressions, still coloured, no doubt, but nearer and nearer to coincidence with pure vibrations?[8] (Bergson 1991, p. 203)

Again, Bergson appeals to our everyday experience to think about the existence of different tensions of duration, appealing to the distortion of our perception of time that takes place in the difference between 'two contemporaneous and distinct persons' that we see in ourselves when we sleep, the first who sleeps for only a few minutes, whilst the second lives in a dream lasting days or weeks (Bergson 1991, p. 207). From this mundane (yet also extraordinary) experience, it does not take too much effort to conceive of the operation of an intuitive faculty that Kant rejected:

> …would not the whole of history be contained in a very short time for a consciousness at a higher degree of tension than our own, which should watch the development of humanity while contracting it, so to speak, into the great phases of its evolution? In short, then, to perceive consists in condensing enormous periods of an infinitely diluted existence into a few more differentiated moments of an intenser life, and in thus summing up a very long history. (Bergson 1991, pp. 207–208)

Thus, against Kant, for whom disorder *precedes* order and the work of intelligence is to impose structure onto the sensible manifold, Bergson advocates an intuitive method that 'seeks to recapture, to get back the movement and rhythm of the composition, to live again creative evolution by being one with it in sympathy' (Bergson 1992, p. 87). One principal way in which Bergson enacts his intuitive method is through the deployment of diagrams to commence the work of speculation on the nature of lived time. In the second chapter of *Matter and Memory*, the 'circuit' diagram depicts the complex of perceptions and memories that typically constitutes even the simplest of our perceptual experiences. Actuality consists, for us, in the duplication of our objects of experience

---

[8]Cf. Bergson's description of the perception of something as 'simple' as red light, that can be disaggregated into billions of vibrations, but that would take more than 250 centuries to perceive if each vibration could be perceived successively (Bergson 1991, pp. 205–206).

with a virtual image of them, on the one side perception, and on the other, memory:

> Each moment of life is split up as and when it is posited. Or rather, it consists in this very splitting, for the present moment, always going forward, fleeting limit between the immediate past which is now no more and the immediate future which is not yet, would be a mere abstraction were it not the moving mirror which continually reflects perception as a memory. (Bergson 1975, p. 165)

This is a critique of the Kantian model of the unity of sensory and conceptual knowledge (the transcendental conditions of possible experience). Instead of a fixed, a priori framework of intuitions (space and time) and concepts through which all conscious knowledge is made possible, Bergson contends that this framework is always generated gradually as we grow, and it continues to mutate over the course of our lives. The a priori/a posteriori distinction was always an approximation of the process frozen at two moments: the point at which a particular level of conceptual awareness is attained and the remaining collections of knowledge accumulated through experience (the height of which is achieved in scientific knowledge). Bergson, instead of reinforcing the a priori/a posteriori distinction, proposes the continuous re-negotiation between virtual and actual engagements with one's world. I am continually making and re-making my relation with the world (or, using Whitehead's terminology, I am continually prehending); and thus perpetually making and re-making my 'self.' This is no longer a traditional concept of the subject: the subject is not constitutive, but rather *creative*. Thus, the circuit sets up the problem of the presentation of a primarily temporal kind of being: the being of perception and memory-formation. The circuit undergoes continual reconstitution in Bergson's destabilization of Kant's constitutive consciousness; the transcendental framework was wrongly envisaged in spatial terms, whereas its activity depicted temporally yields the moving image of a circuit.

3 The Significance of Lived Time

How might we envisage this 'circuit' in terms of the plural times of relativity? The endeavour to rediscover duration through an intuitive inspection of, first, one's own experience, and then to an analysis of other processes, does not cease at the door of theoretical physics. This is the contention of *Duration and Simultaneity*. Bergson refers back to his hypothesis of a single, physical time as set out above, repeating his idea of 'a duration of the universe', or 'an impersonal consciousness that is the link among all individual consciousnesses, as between these consciousnesses and the rest of nature' (Bergson 1999, p. 31). There are two ways of thinking about the relation between individuals and time. Let us think about them using Bergson's circuit diagram from *Matter and Memory*.

Firstly, a single consciousness with its own duration is capable of encompassing certain aspects of its immediate environment and it is, in fact, what constitutes the barest of perceptual experiences. This is indicated by the circuit AO in the diagram. Now, diverging from the original diagram, consider the synchronization of durations that this basic interaction requires. My time and the time of my object must be capable of 'taking account of' one another (this would echo the rudimentary connectedness indicated in Whiteheadian 'prehensions'). The synchronization of two or more durations can thus be depicted as the addition of concentric circles around the subject or centre of action. Now the image, discussed earlier, of the ever greater tensions of consciousness extended so far out as to encompass the whole universe (implying a single Time or Duration) allows us to imagine the possibility of a vast, interconnected web of 'images' that constitute the material universe (both epistemologically and ontologically, in concert with other process ontologies or cosmologies).[9] Then, withdraw again to the single consciousness and its

[9]The idea of a single time seems, at first glance, to be at odds with Bergson's own statements on the interactions between different durations (higher and lower tensions) of life in *Creative Evolution*. However, Bergson repeats what he says in that earlier text concerning the specificity and contingency of the configurations of life that have evolved on this planet. Life could have evolved otherwise, and expressed its tendency in ways other than the proliferation of individuals along divergent lines of evolution. Consciousness has, in fact, become diffused and distributed across effectively distinct individuals. Alternative 'tensions' of consciousness remain, in theory, possible.

immediate environment (depicted, say, by only two concentric circles). We can add another consciousness into the diagram, depicted as a third concentric circle. This is to represent the possible interactions and perceptions of one consciousness by another. The spaces in between the two consciousnesses indicate their relative proximity or distance to any chosen degree. Regardless of how far away they are from one another, the intervening spaces (durations) must be capable of synchronization with both of their own durations.

Now consider these relations in terms of the claims of Special Relativity. Two clocks are synchronized and then separated; one remains stationary and the other moves at speed. According to relativity theory, the motion of the second clock will cause a slight material contraction in space and a slight dilation in time. For the sake of illustration, say that the clock in motion proceeds twenty days; the second stationary clock two hundred days. The clocks are, of course, read by observers travelling along with them. Interpreted through our diagrams (the concentric circles), SR suggests that the two consciousnesses must be isolated; that is, separated from one another by an infinite difference in time; that plural times in relativity coexist in spacetime in complete isolation from one another; that time *is nothing*, because *it does nothing* (to echo Bergson's critique of certain accounts of evolutionary time in *Creative Evolution*). What is at work here, of course, is a fundamental incompatibility between *metaphysical time* according to which 'it is impossible to imagine or conceive a connecting link between the before and the after without an element of memory and, consequently, of consciousness. […] because duration is essentially a continuation of what no longer exists into what does exist, [a] real time, perceived and lived' (Bergson 1999, p. 33), and the combination of *physical time* (real time contraction/dilation) with *psychological time* (giving rise to imagined contractions/dilations of the experience of time). For Bergson, time is only conceivable together with some degree of consciousness (by which he means the most rudimentary activity: the connecting of before and after). The critique of plural times in SR thus also raises the very particular problem of the relationship between science and metaphysics: either they are complementary, or they become dogmatic. This incompatibility exists even more starkly in the assumption, from the psycho-physical perspective, that both observers would experience the passage of time *at the same speed*, whilst the physical reality is one of plural time-speeds.

With Bergson, if we extend the hypothesis (Bergson 1999, p. 32) of the living individual subject who can assume a harmony between her inner time and that of her immediate surroundings (and how this would ultimately overlap with the inner time and corresponding outer time of another subject some distance away), then we seem to require some reference or assumption of a single universal time. At what point, we might ask of the physicist, does the inner time of the subject, the time of the subject's environment, and that of the environment's exterior start to diverge? Bergson thus claims to

> demonstrate in more concrete fashion, the increasing distortion of the surface image and the unchanging preservation of its inner relations as the speed is adjudged increasing. We shall thus catch, as if on the wing, the genesis of the plurality of times in the theory of relativity. We shall see its meaning take physical shape before our eyes. And, at the same time, we shall extricate certain postulates which this theory implies. (Bergson 1999, p. 86)

In SR, the reciprocity of motion assumes that the positions of the two clocks are interchangeable as systems of reference. The Lorentz equations suggest that system 1 (S) will advance two years, and system 2 (S') will advance two hundred years. But if there is reciprocal motion (either S or S' can stand as reference point), then in both cases only two years must have advanced, because both systems experienced by two different individuals involve the identical experience of time elapsed. Either motion is reciprocal (Cartesian) or there is one system of reference (More). In relativity, there is only psychological time and physical time. The former refers to the experience that can be distorted by emotional states (boredom, impatience, excitement, etc.); the latter is actually distorted at speed and in the proximity of strong gravity. Bergson's fundamental claim about the reality of lived time is either accepted, and it resolves relativity into a single time and imagined multiples of it, *or* there is no metaphysical time (as Einstein states).

Time, as indicated through inspection of our own consciousnesses, is multiple in the sense that memories are numerous but confused in their virtual condition, but they can be delimited and enumerated through the process of actualization into real movement. The hypothesis of different durations (as the theory of life in *Creative Evolution* posits) is an exploration of different lived experiences of time: durations as higher and lower tensions of consciousness, where some can encapsulate greater concentrations of information in a single moment than others, constituting a lattice of multiple durations. As Bergson says, however, he sees no reason to extend this to the physical universe in general. Historically, there are ways in which this can be done, such as through a panpsychist approach,[10] but Bergson's objection to the plurality of physical time concerns their collective status as 'real.' Only 'lived time' is real.

To summarize, the physicist appeals not only to his own perception of his system, but to 'all the mental views ascribable to all the physicists perceiving that system in motion at every possible speed' (Bergson 1999, p. 86). What Bergson's analysis highlights is (a) the relation between physical, psychological, and metaphysical duration (the third denied by Einstein) and (b) how the means of measurement (the clock, of whatever nature or degree of accuracy) becomes indistinguishable from time itself; the latter is reduced to the former:

> The real observer inside his system is indeed aware of both the difference between, and the identity of, these two different times. He lives a psychological time, and, with his time, all the more or less expanded mathematical times merge; for… in the measure that he mentally accelerates the motion of his system—the light-lines lengthen, but they all fill the same lived duration. (Bergson 1999, p. 89)

Hence relativity confirms a single time, even as it proposes multiple times. The observer elides psychological time with the light-line in order to think about her system in motion; and then says that her light-line (thought now as time) has lengthened. However, once she really occupies that other system, she cannot really experience a lengthening of time.

---

[10]I will explore this further in the next two chapters.

She can only experience her own psychological duration (Bergson 1999, pp. 89–90):

> And, as it is of the essence of physics to identify the thing with its measurement, the 'light-line' becomes both the means of measuring time and time itself. But then, since the light-line elongates, while remaining itself, when we imagine as in motion yet leave at rest the system in which it is observed, we shall obtain multiple, equivalent times; and the hypothesis of the plurality of times, characteristic of the theory of relativity, will appear as conditioning the general evolution of physics as well. Times thus defined will indeed be physical times. They will be only conceived times, however, all except one, which will actually be perceived. The latter always the same, is the time of common sense. (Bergson 1999, p. 93)

If SR reduces time to the means of measuring it, then we require clarification of Bergson's alternative to this reduction. To do this, the next section explores some of the affinities between methods at work in the use of diagrams and images in both Bergson's own work and in the pedagogy of General Relativity exemplified in the visualization of black holes.

4 Multi-dimensional Thinking

Bergson uses the image of a shooting star to illustrate the difference between unfolding duration and the idea of an unrolling, measurable time that effectively obscures the unfolded space that serves as its foundation. The 'fiery path' traced out by a shooting star gives us *both* a line that is divisible at will and the 'indivisible mobility that it subtends' (Bergson 1999, p. 34). However, when we mistake the resulting line or path for the movement itself, then we have mistaken an unfolded space for the original unfolding duration that preceded it. So too with the movement of evolution (*élan vital*) and the evolved successive generations of living organisms that have resulted from it; the physical act, and the will that drove it. Let us think about the retrograde movement of truth now in terms of the multi-dimensional representations of space and time. 'Spatialized time' is a fourth dimension of space; its addition allows the successions in space in three dimensions to be perceived as simultaneities in the fourth (as we move from point to line to cube, etc.):

> Whether a universe has three, two, or a single dimension, or even none at all and reduces to a point, we can always convert the indefinite succession of all its events into instantaneous or eternal juxtaposition by the sole act of granting it an additional dimension. (Bergson 1999, p. 41)

This exposes the problem of viewing time as a flat plane: the events of a life (and by extension all things) can be laid out on a plane, but we tend to forget this is only for the purposes of reflection. A fourth dimension added to the third, *as if* it is another dimension of space, generates the retrograde truth that a life *can* be summed up in this way. The original duration is reduced to the unfolded, spatial, remnants of it, and is then retroactively superimposed onto this unfolded space to become an 'unrolling' succession of moments:

> Immanent in our measurement of time, therefore, is the tendency to empty its content into a space of four dimensions in which past, present, and future are juxtaposed or superimposed for all eternity. This tendency simply expresses our inability mathematically to translate time itself, our need to replace it, in order to measure it, by simultaneities which we count. These simultaneities are instantaneities; they do not partake of the nature of real time; they do not endure. They are purely mental views that stake out conscious duration and real motion with virtual stops, using for this purpose the mathematical point that has been carried over from space to time. (Bergson 1999, p. 42)

How, then, do we move from an *unfolded* space and *unrolling* time (of pre-formed elements into a new configuration) to an *unfolding* duration (in which the new parts are not contained, but are generated by, the preceding)? One immediate suggestion is in the experience of drawing[11]; such an experience perceived from within is an *unfolding* of duration,

[11] Cf. Gemma Anderson's work cited in the previous chapter.

whereas the line traced from point to point on the paper is the *unfolded* succession (Bergson 1999, p. 34). I want to suggest in the rest of this section that further explorations of different methods of visualizing processes can contribute to Bergson's intuitive method for thinking about process as unfolding duration, together with contemporary physical speculations on the implications of General Relativity for an understanding of reality as multi-dimensional.[12] I return here to the diagrammatic representation of a black hole, and the curvature of spacetime that it effects, with which I commenced this chapter, and its efficacy as a visual analogue of Bergson's theory of the subject as centre of action.

Special Relativity recommended a radical shift away from Newton's model of interacting material bodies within an absolute space, an absolute time, an aether, and the relative gravitational effects of those bodies on its surroundings.[13] Einstein's generalization of relativity (1912) to include gravity incorporates Hermann Minkowski's proposed four-dimensional model of the universe (circa 1908) (Thorne 1994, pp. 107–108), and this will take us to the final aspect of Bergson's critique of Relativity as set out in *Duration and Simultaneity*. Whilst the equivalence principle allows for the interchangeability of descriptions of the relativistic effects across multiple 'reference frames' (e.g. an observer sitting in the cabin of an accelerating rocket, at a certain speed, may feel the equivalent effects of the gravitational pull of the Earth: 1G; or multiples of that effect: 2G, 3G, etc. Both reference frames can be treated as equivalent as far as the rules of Special Relativity are concerned: spatial contraction, time dilation, etc.), the incorporation of a model of spacetime as a four-dimensional field allows for a more tangible alternative image of the universe to the Newtonian one. Instead of bodies interacting in absolute space and time, spacetime is a fabric that is susceptible to distortions,

---

[12]Cf. The difference between fern leaf growth (unrolling movement) and other plants (emergent growth) in the previous chapter to *visually* illustrate unrolling versus unfolding (this provided Goethe with a means to affirm his theory of the protean leaf, but the fern's growth proved the exception rather than the rule. The instructive distinction is between the cinematograph and a living organism.

[13]Koyré (1968) notes that Newton was keen to distance himself from any idea of an 'innate gravity' possessed by matter; pointing instead to the sufficiency of assuming varying gravitational effects to explain certain phenomena, without necessarily having to name the source of that gravity (Koyré 1968, pp. 178–179).

warping, or curvature, depending on the mass of the bodies that subsist within it. After Einstein, *gravity* will no longer be thought as a force (whether emanating from matter or some other source), but rather the after-effects of spacetime curvature, replacing or renaming what Newton referred to as tidal gravity, or the tendency of things to fall towards a body's centre of gravity (a planet's tidal gravity will be perceptible as a set of angled descents of any bodies falling towards the planet's surface converging towards a point). Spacetime curvature, then, will help to explain phenomena such as black holes, gravitational lensing, and gravitational waves (Thorne 1994, p. 92).

As we saw, Bergson's critique of SR includes a criticism of the application of additional dimensions to our perceived three-dimensional reality as nothing more than the spatialization of time (in the case where time is assumed to be a fourth dimension): 'The thing is what is perceived; the expression is what the mind puts in place of the thing to make it amenable to calculation' (Bergson 1999, p. 97). The assumption of time as the fourth dimension of space views each system in the plurality of times as equal and interchangeable. I do not want to suggest that the development of Special Relativity into General Relativity solves Bergson's problems with relativity, but the later theory does facilitate more productive ways of deploying multi-dimensional visualizations as aids to thinking about the nature of process (in its ontological sense). What might be called the 'regressive' element of SR, from a Bergsonian stance, is retained in GR insofar as it reiterates the privileged position of the physicist in the application of 'reference frames' to represent the plural times of relativity. However, there is a more 'productive' element to this theory in the contemporary use of multi-dimensional visualizations that allow for an engagement with Bergson's model of memory (the cone diagram) on a more equal footing.

In a passage from his 1966 book, *Bergsonism*, Deleuze argues that 'Bergson does not use the word "unconscious" to denote a psychological reality outside consciousness, but to denote a nonpsychological reality—being as it is in itself. Strictly speaking, the psychological is the present. Only the present is "psychological;" but the past is pure ontology; pure recollection has only ontological significance' (Deleuze 1991, p. 56). The core problem Deleuze is addressing here is that of the 'location' of memories if, as Bergson states, they are not contained in the brain. 'Where' indeed are memories stored if not in the brain? Of course the question, as Deleuze's reminds us, is born of a badly analysed composite. It is a false problem, based on an immense metaphysical presupposition about the nature of matter and mind, and the relation between the two.[14] The possible configurations of the mind-matter relation are well known, but the questions that arise if we follow Bergson's particular claims about it through to its consequences are the focus of the remainder of this chapter.

[14]Walter Elsasser (1998) in his attempts to unify quantum mechanical principles with biological research, refers to Bergson's framing of the mind-brain relation as an exemplary formulation of the problem of biological reductionism. When Elsasser cites Bergson's *Matter and Memory* for its formulation of the problem of memory, it is to clarify the distinction between 'physical' and 'biological' forms of memory, or homogeneous replication and heterogeneous reproduction. As Elsasser underlines, 'There is no evidence that would indicate the existence of a mechanism of transcription whereby contents of cerebral memory would end up by becoming sequences of a long molecular polymer after the manner of the genes, together with a reverse 'readout' mechanism that would transform molecular sequences again into memory contents: (Elsasser 1998, p. 87) The difference between a concept of cerebral matter as an identifiable storage device (a mechanistic construct) and a distinct form of 'cerebral memory' is, as Elsasser acknowledges, one that is discussed at length by Bergson. For example, in the second chapter of *Matter and Memory,* Bergson defines this problem in terms of a difference between 'recollections strung out along the course of time' and movements that are nascent actions in space (Bergson 1991, p. 79). We have here, now in Bergson's use of the terms, the distinction between the relative heterogeneity of the complex multiplicity of memories and the homogeneity of distinct perceptions and actions. In a particularly evocative passage, Bergson describes the process of recollection as a progressive narrowing of the intricate detail of memories 'in dream' into less complicated 'generalizations' more readily applicable to a wide range of living situations: 'to act is just to induce this memory to shrink, or rather to become thinned or sharpened, so that it presents nothing thicker than the edge of a blade to actual experience' (Bergson 1991, p. 106). Memory needs to be described less as a repository of past experiences, and more as a pure process of 'contraction and expansion' depending on the imminence of an organism's actions. See also Stephen Robbins (2006) who has advocated a return to Bergson in neuroscientific research for alternative definitions of perception. (2006): 'By the "photograph developed in the very heart of things and at all points in space," Bergson was envisioning the matter-field as a dynamic interference pattern—a holographic field. This was 50 years before Gabor discovered the principles of holography' (Robbins 2006, p. 367).

In the last chapter, I pointed to aspects of Bergson's philosophy that might serve as a critical foundation for thinking about process using different imaging techniques. Bergson's critique of the cinematographical mechanism of thought cautioned against eliding a convenient technological trick such as still photography with the process that it disassembles. Secondly, Bergson's suggestions for developing a philosophical intuition entailed a dual focus on *method* as well as *object* as a means of turning our attention away from action alone to a kind of contemplative action. Understanding *how* we tend to think about temporal phenomena (consciousness, life) will help us to learn more about *what* these phenomena are. Diagrams and drawings can serve as heuristic tools, but the experience of the *limitations* of trying to represent a living process visually can itself be a better heuristic tool than any static diagram. Once the thought is set down in words or images, its activity has already passed beyond our grasp. What use, then, can we make of Bergson's diagrams for the discussion so far?

We saw that Bergson, in *Creative Evolution*, develops further his speculative endeavour to think in terms of time rather than in terms of space. This text is both a study of the metaphysical underpinnings of 'Neo-Darwinian' evolutionary theory in a number of Bergson's contemporaries in the late nineteenth century, and a contextualization of the evolution of human intelligence within a broader development of conscious 'tendencies' that is discernible across species lines. Human consciousness is not to be understood as the emergence of an exceptional form of existence in evolutionary history. It is, rather, one example of an organic tendency towards the manipulation of matter, alternatives to which are found in animals of varying complexity all the way up from the amoeba to the orangutan. Consciousness, in what I would call the de-anthropomorphized sense I outlined earlier using the circuit diagram, is merely specialized in multiple forms across multiple species. Once we consider each consciousness no longer as a specific 'subject' in the Kantian sense, we build up a picture of material reality as a multiplicity of 'tendencies' or 'vectors' or 'processes.' Individual consciousness, that thinking, creating, vital force, is thus an immediately accessible indicator of the nature of temporal reality. The microscopic activity of each individual consciousness simply recapitulates the macroscopic activity of the evolution of organic life considered as a whole.

Example 1: The Cone Diagram

Whereas the *circuit* diagram sets up the fundamental dynamism of Bergson's theory of memory, the *cone* diagram attempts to present the movement between different 'states' of consciousness (again, the identification of states wrongly locating mind in regions of space), now conceived as different 'intensities' or 'tensions' of consciousness. The planes of the cone, not to be confused with the spatial locations of states of mind, are not themselves the centre or subject of the diagram but are subordinate to the process in which we pass through the different tensions of consciousness. Bergson describes the different elements of the diagram as follows:

> If I represent by a cone SAB, the totality of the recollections accumulated in my memory, the base AB, situated in the past, remains motionless, while the summit S, which indicates at all times my present, moves forward unceasingly, and unceasingly also touches the moving plane P of my actual representation of the universe. At S, the image of the body is concentrated, and, since it belongs to the plane P, this image does but receive and restore actions emanating from all the images of which the plane is composed. (Bergson 1991, p. 152)

If we take the plane P as presented in the cone diagram (that is, the plane of action) within the theory of plural times or systems, the plane P is effectively replicated and 'stacked' so that each new plane P adds to form a geometrical 'patchwork' that in turn constitutes the material universe understood as spacetime. Bergson's contention is that the physicist of SR assumes that one can pass form one plane to the next, with each plane being stretched or squeezed according to its motion and speed (and with the additional claim that within each system, the individual's *experience* of time will remain constant). The problem is that once one moves from one plane to the next, the plane one now occupies or *lives* is *really* experienced, whilst the others become only *virtual* or *phantasmal.* One can only range through a universe made up of these 'stacked' planes of action as a result of a retrospective judgement, that is, once the real-time experience (like the fiery path of the shooting star in the making) has been 'unmade' and then reconstructed. Real time or duration in the making only happens once, and is only experienced once. It is formed as it is lived. To recount its process is to retrace its steps after its 'death.' So much for a retrogressive reading of the cone diagram. In order to engage with Bergson's formulations of the nature of time (through memory) in conversation with physical formulations of time, I will turn to the 'embedding diagrams' used in the pedagogy of relativistic physics.

We are unable to perceive things beyond one, two, and three dimensions. Typically, time is conceived as a fourth dimension (via Minkowski), but in the physical descriptions of the curvature of spacetime up to seven dimensions are conceived. We have seen that Bergson objects to the addition of spatial dimensions as an effective way of thinking about time. However, for the purposes of reconciling the problems of multi-dimensional spacetime and Bergson's intuitive method for thinking about duration, I will first outline the ways in which models deploying four or more dimensions aid physical research. Thinking in terms of merely a fourth dimension of space (here, conceived separately from a fourth dimension of time) is problematic, but it also solves certain problems in the description of spacetime curvature. If gravity is no longer conceived as a force 'belonging' in some way to objects that pulls other objects towards it, and it is now reconfigured as curvatures in the fabric of spacetime itself, then we are presented with a new problem: *into what* does spacetime curve? There are several elements that are important for understanding this image: (i) spacetime is not a container in which objects affect one another through the action of certain forces; (ii) spacetime is a fabric weaving together all parts of the universe; (iii) objects with higher densities cause warping or curvature of spacetime itself; (iv) we observe this curvature as gravitational pull; (v) the Earth exerts a certain pull by virtue of the curvature its presence effects in spacetime; (vi) if there is curvature, it must *curve into* a further dimension of space (e.g. a fourth or fifth dimension, etc.).

Example 2: Embedding Diagrams

In order to think about these additional dimensions, the use of 'embedding diagrams' has become common practice. Indeed, popular scientific descriptions of phenomena such as black holes and worm holes typically deploy embedding diagrams giving us the familiar 'funnel' image of the black hole. The term 'black hole' reinforces this type of image, but the funnel is only an embedding diagram in which one or more dimensions has been removed in order to present a more easily digestible two/three-dimensional proxy for a phenomenon that exists in four or more dimensions. The thing in itself requires a much more complex description. The following only offers some tentative steps in that direction.

In 1784, the Reverend John Michell in a correspondence with Henry Cavendish, postulated the existence of stars so massive that it would be impossible for even light to escape their gravitational pull. The prediction predates considerably the modern description of black holes, describing the possibility that observation of distant clusters of visible stars might yield knowledge of the presence of less luminous stars, ones that are 'darkened' by the effects of their superior mass:

> from the infinite variety which we find in the works of the creation, it is not unreasonable to suspect, that very possibly some of the fixed stars may have so little natural brightness in proportion to their magnitude, as to admit of their diameters having some sensible apparent size, when they shall come to be more carefully examined, and with larger and better telescopes than have been hitherto in common use.[15] (Michell 1784, p. 48)

Michell's prediction, later also made by Pierre Simon Laplace in 1799, presents a number of the elements by which we now understand black holes (Thorne 1994, pp. 122–123). For example, the escape velocity in relation to any planet or star is the speed required for light particles to break free from the gravitational pull of that body (calculated in proportion to the square root of the mass of the star or planet divided by its

[15]See Michell's precise explanation for this lack of natural brightness: 'Hence, according to article 10, if the semi-diameter of a sphaere of the same density with the sun were to exceed that of the sun in the proportion of 500 to 1, a body falling from an infinite height towards it, would have acquired at its surface a greater velocity than that of light, and consequently, supposing light to be attracted by the same force in proportion to its vis inertiae, with other bodies, all light emitted from such a body would be made to return towards it, by its own proper gravity' (Michell 1784, p. 42).

circumference). The smaller the circumference of the star in relation to its mass (e.g. a dense star whose surface is closer to its centre, and thus possessing a stronger surface gravity), the greater the escape velocity must be. The velocity required to escape its gravity must be equal to or faster than the speed of light. The 'horizon' of a black hole is the farthest limit of its gravity. Beyond the horizon, towards the singularity at its heart, it is impossible for light to escape and it thus remains hidden to us. The existence of black holes has come to be accepted as part of the story of the life cycle of stars. Stars age and then collapse into white dwarves, neutron stars or black holes, each of these three types of body possessing the same mass, whilst a black hole does not 'contain' matter. The star's matter has imploded such that it is now composed of warped spacetime[16] (Thorne 2014, p. 22).

How, then, can we think this curvature or warpage, without matter, in the fabric of spacetime itself? An embedding diagram itself makes use of an analogy. If we were to imagine a world populated by two-dimensional beings, then, for them, our three-dimensional universe would be experienced as a flat plane. A three-dimensional object, such as a sphere is effectively an 'impossible object'. To a 2D being, a sphere might look like a series of flat circles of first increasing and then decreasing size. But from these snapshots, the 2D perceiver could derive secondary knowledge of the 3D curvature of the object.[17] For the purposes of communicating an imperceptible fourth dimension, the key components of the black hole are thus presented in an embedding diagram in a digestible 2D/3D form that serves as a proxy for a 3D/4D phenomenon.

Example 3: Visual Effects

The insights derived from this kind of presentation are expanded and tested in visualizations of what a black hole might look like from the perspective of a human observer positioned near to it in Kip Thorne's

[16]Black holes are thought to originate either from collapsed/imploded stars or they become supermassive black holes made up of smaller black holes formed in turn from collapsed stars.

[17]Thorne (1994) goes into more detail about the tracing of two parallel lines across a flat plane, as opposed to the same parallel lines along a spherically curved surface. In the former case, the lines will never cross. In the latter case, the lines will cross, thereby indicating that there is curvature. Physicists are able to visualize predicted phenomena through a similar deployment of 2D/3D proxies for 4D (or higher) objects.

collaboration on the visual effects for the film *Interstellar.*[18] From the perspective of a human observer, the experience of the warpage of space in three dimensions, rather than just the 2D/3D rendering in the form of a funnel, would be manifested in the form of a *sphere.* The singularity, which is the result of the imploded matter of the former star and that now exists in its place (and it is, the speculation goes, a place where space and time are infinitely warped; where relativistic and quantum laws perhaps meet), remains invisible to us. What we would see, rather is the warping space around it. Now this space warping, as Thorne points out, is not produced *by* the singularity, but by the warping itself.[19] Thus, what we would see is not a 'hole' at all, but the curved space casting a shadow on the field of stars that we would otherwise see (Michell's suggestion of it as a 'dark star' thus seems appropriate). Importantly, the shadow implicates the perspective of a human observer into the visualization of the predicted or theoretically possible phenomenon of a black hole and its influence on surrounding spacetime.[20]

Take the example of gravitational lensing which occurs around our Sun, but also around any other object of significant enough mass. In the film, *Interstellar*, the phenomenon of gravitational lensing is used in the visualization of the black hole (named 'Gargantua'), presenting us with a more 'veritable' image of a 'dark star'. Gargantua is a spinning black hole, the centrifugal motions of which attract surrounding objects and debris to it, creating the accretion disk around it (much like the rings of Saturn). The double 'corona' we can see around Gargantua is a depiction of gravitational lensing. The mechanism referred to as the gravitational slingshot (which has been practically applied, using the gravity of planets) is capable of deflecting or accelerating objects in motion. Now this deflection/acceleration also affects light rays, and this is what gives us gravitational lensing. In the final shot, two images are created as a result

[18]See James et al. (2015).

[19]'The warping energy is stored in the black hole's warped space. And for a black hole, that energy of warping is so great that it generates the warping' (Thorne 2014, p. 46).

[20]Maurice Merleau-Ponty's writings note the embodiment of the human perspective in the paintings of Cezanne. The speculations on the appearance of a black hole for a human observer in close proximity to it seems to echo, if unintentionally, this drive to see things from a human/organic/embodied perspective.

of gravitational lensing. Any light that manages to escape the pull of the black hole's gravity continues to travel in a straight line. What we see is light from the accretion disk that is maintained in orbit around the black hole since it is far enough from Gargantua's horizon (the 'critical circumference' of the black hole) to either dwell or escape. Whatever is behind the black hole (in relation to a human observer on our side of it) is still visible to us. Thus the sphere swallows some light, but the rest is 'lensed' around it, giving us the doubled image of both the star field behind it and the accretion disk (Thorne 2014, pp. 94–97).

5 Summary

What this brief survey of visual tools demonstrates first is that the complexity of interactions, described in Example 3, between the enfolded spacetime and the light that surrounds it, or is absorbed by it, is belied by a simple embedding diagram (Example 2) depicting the black hole as a 'funnel.' The depth of the funnel indicates only 'a certain aspect of a Schwarzschild black hole, namely the intrinsic geometry of a two-dimensional surface in the space around the black hole'[21] (Marolf 1999, p. 919). Taking a broader view in terms of method, both Thorne and Bergson emphasize and prioritize the exploration of the limits of thought. For example, Thorne notes the ways in which Newtonian and then Einsteinian laws are always being pushed to the edges of their *domains of validity* in the formulation of new experiments. Of his own analysis of the history of modern physics, Thorne writes:

> the process repeats. We shall meet the repetition in coming chapters: The failure of special relativity when gravity becomes important, and its

[21]Marolf's article details an attempt to produce a diagram that will reflect the temporal aspects of a black hole's behaviour: 'To make our new kind of diagram, we will again choose a two-dimensional surface and embed this surface in a three-dimensional space. However, since we wish to show the space*time* aspects of the geometry, our new surface will include both a spacelike direction and a timelike direction. That is, some pairs of events on our surface will be separated by a timelike interval while other pairs will be separated by a spacelike interval. We therefore need to draw the surface inside three-dimensional (2 + 1) Minkowski space as opposed to Euclidean three-space' (Marolf 1999, p. 921).

> replacement by a new set of laws called *general relativity* [...]; the failure of general relativity near the singularity inside a black hole, and its replacement by a new set of laws called *quantum gravity*.' (Thorne 1994, pp. 84–85)

The labour of the physicist is described in terms of an intuitive grasp of how things might work, or 'approximations,' as an important way of working at the edges of domains of validity, and essential for the furthering of scientific knowledge. Echoing in some ways the 'regulative' use of ideas, underpinned by the assumption that there is an ultimate set of laws that forces the universe to behave in certain ways, Thorne opts to avoid the term 'theory' in favour of 'physical laws.' An adoption of the latter phrase, he continues, is meant to indicate the underpinning assumption, whilst also allowing for changes that might be necessary at the edges of domains of validity. In this, I see no disagreement with Bergson's approach to intuition. It is also worth comparing Bergson's stance on 'lines of fact' in which we deal with probabilities (cf. 'approximations' in Thorne) as a means of furthering knowledge at the edges of 'the known.'

The pedagogical value of the subtraction of one or more dimensions from the diagrammatic model of the black hole can, I would argue, be replicated in the use and interpretation of Bergson's diagrams, and thus guide us towards a more considered appreciation of the virtual-actual relation depicted in the images of the circuit and the cone. Such experimentation with multi-dimensional images allows for the conception of realities unreachable by ordinary perception (or at the edges of ordinary perception), which is compatible with a Bergsonian view of intuitive thinking (at the edges of intellectual activity). Instead of making recourse to a dualistic view of a virtual realm and an actual realm, we can now view the cone diagram as a 2D/3D proxy for a 3D/4D process, again repeating the intent of the regulative use of ideas, whilst bearing in mind the retrograde movement of truth. The usefulness of methods for thinking across dimensions should not be 'reified' and thus mistaken for a true grasp of reality. Viewing Bergson's cone diagram in this way has a number of implications. (i) It provides a way to address the error of viewing the plane of action as a 2D plane into which a 3D cone of memory is inserted. The error promotes the illusion of a spatial realm of 'the virtual' interposing upon our present. It reinforces, as we saw above, the idea that multiple times could be stacked or patch-worked together to form a three-dimensional space (or spacetime). The difficulty of thinking the virtual/actual relation is analogous to the difficulty of thinking the 4D/3D relation. (ii) With diagrammatic representations, it is always important to remember that we are dealing with proxies, and that we must guard against the over-interpretation of the spatial organization of ideas. (iii) There is no sense of a priority of virtual over actual, not even in the idea of an ontological priority of memory or duration/time. If memory is said to be prior to perception, it is first a provocative statement that is intended to disrupt the usual order of things in typical accounts of the memory-perception-action relation: it is *not* the case that memory is merely an after-effect or a weakened perception. (iv) Memory (Duration) is not a container or repository of memories (or the past). The past coexists virtually with the present. The coexistence and co-integrity of multiple dimensions that we can demonstrate by thinking about the relation between two-dimensional and three-dimensional objects (and the projected integration of third and fourth dimensional objects) helps us to think about the relation between memory and perception, virtual and actual. (v) The conscious subject, understood as centre of action, exerts a certain distorting influence on its surroundings, and the material universe as aggregate of images is a 'fabric' that is capable of suffering distortions depending on the tensions of memory (the concentration or contraction of memory) of the centre of action, be it human or more-than-human.[22]

[22]The neuroscientific study of our ability to recognize faces can help to demonstrate the model of memory depicted in the cone diagram. We might also consider the problem in relation to pathological cases such as facial recognition disorder. Diagnostic descriptions of the normal and pathological operation of facial recognition have indicated a distinction from object recognition:

> According to the CPH [configural processing hypothesis], the visual system contains no components that are specialized for processing faces. It does, however, have procedures specialized for configural processing: procedures that allow the recognition of an item as a whole in one glance. Tests of configural processing require identification of objects with deleted portions so that the perceiver must combine a number of individually meaningless parts to form a structured percept. According to CPH, prosopagnosia results from a loss of configural processing. In consequence, it predicts that all prosopagnosics should be severely impaired in tests of visual closure. (Duchaine 2000)

References

Bergson, Henri. 1975. *Mind-Energy: Lectures and Essays*, trans. H. Wildon Carr. Westport, CT: Greenwood Press.

Bergson, Henri. 1991. *Matter and Memory*, trans. N.M. Paul and W.S. Palmer. New York: Zone Books.

Bergson, Henri. 1992. *The Creative Mind: An Introduction to Metaphysics*, trans. Mabelle L. Andison. New York: Citadel Press.

Bergson, Henri. 1999. *Duration and Simultaneity*, trans. Leon Jacobson. London: Clinamen.

Berkeley, George. 1975. *Philosophical Works*. London: Everyman.

Deleuze, Gilles. 1991. *Bergsonism*, trans. Hugh Tomlinson and Barbara Habberjam. New York: Zone Books.

Duchaine, Bradley C. 2000. Developmental Prosopagnosia with Normal Configural Processing. *Neuroreport* 11 (1): 79–83.

Elsasser, Walter M. 1998. *Reflections on a Theory of Organisms: Holism in Biology*. Baltimore: Johns Hopkins University Press.

Gunter, Pete A.Y. 2008. Bergson's Divided Line and Minkowski's Psychiatry: The Way Down. *Chromatikon: Annales de la philosophie en procès/Yearbook of Philosophy in Process* 4: 107–119. https://doi.org/10.5840/chromatikon2008410.

The above refers to developmental prosopagnosia, as opposed to acquired prosopagnosia (occurring as a result of stroke or injury) which is famously referenced in Oliver Sacks' essay on 'The Man Who Mistook His Wife for a Hat' (Sacks 1985). After 'confabulating non-existent features' that the patient, Dr. P., 'sees' by not quite managing to look properly at a cover of *The National Geographic*, Sacks recounts:

> I must have looked aghast, but he seemed to think he had done rather well. There was a hint of a smile on his face. He also appeared to have decided that the examination was over and started to look around for his hat. He reached out his hand and took hold of his wife's head, tried to lift it off, to put it on. He had apparently mistaken his wife for a hat! His wife looked as if she was used to such things. (Sacks 1985, pp. 10–11)

The suggested complexity of the disorder opens up some potentially interesting 'temporal' accounts of its dynamic: the distinction between face and object-oriented recognition points to different dynamics of learning at play. Furthermore, as Pete Gunter outlines in his essay 'Bergson's Divided Line and Minkowski's Psychiatry: The Way Down', some pathological phenomena can be articulated in terms of the difference between the vital and material orders. I would suggest that this explanation can be extended to cases in which the recognition of faces fails (as it often does in my own case). That is, one falls into a kind of material repetition, at best only able to discern the rudimentary form of a face and, at worst, unable to distinguish between faces and inanimate objects.

James, Oliver, Eugénie von Tunzelmann, Paul Franklin, and Kip S. Thorne. 2015. Gravitational Lensing by Spinning Black Holes in Astrophysics, and in the Movie Interstellar. *Classical and Quantum Gravity* 32 (6): 065001. https://doi.org/10.1088/0264-9381/32/6/065001.

Kant, Immanuel. 1929. *Critique of Pure Reason*, trans. Norman Kemp Smith. London: Macmillan.

Koyré, Alexandre. 1968. *From Closed World to Infinite Universe*. Baltimore: John Hopkins University Press.

Lorentz, H. A., A. Einstein, H. Minkowski, and H. Weyl. 1952. *The Principle of Relativity: A Collection of Original Memoirs on the Special and General Theory of Relativity*. New York: Dover.

Marolf, Donald. 1999. Spacetime Embedding Diagrams for Black Holes. *General Relativity and Gravitation* 31 (6): 919–944. https://doi.org/10.1023/A:1026646507201.

Michell, John. 1784. VII. On the Means of Discovering the Distance, Magnitude, & c. of the Fixed Stars, in Consequence of the Diminution of the Velocity of Their Light, in Case Such a Diminution Should be Found to Take Place in Any of Them, and Such Other Data Should be Procured from Observations, as Would be Farther Necessary for that Purpose. By the Rev. John Michell, BDFRS In a letter to Henry Cavendish, Esq. FRS and AS. *Philosophical Transactions of the Royal Society of London* 74: 35–57. https://doi.org/10.1098/rstl.1784.0008.

Olma, Sebastian. 2007. Physical Bergsonism and the Worldliness of Time. *Theory, Culture & Society* 24 (6): 123–137. https://doi.org/10.1177/0263276407078715.

Robbins, Stephen E. 2006. Bergson and the Holographic Theory of Mind. *Phenomenology and the Cognitive Sciences* 5 (3–4): 365–394. https://doi.org/10.1007/s11097-006-9023-1.

Sacks, Oliver. 1985. *The Man Who Mistook His Wife for a Hat*. London: Duckworth.

Scott, David. 2006. The 'Concept of Time' and the 'Being of the Clock': Bergson, Einstein, Heidegger, and the Interrogation of the Temporality of Modernism. *Continental Philosophy Review* 39 (2): 183–213. https://doi.org/10.1007/s11007-006-9023-4, p. 189.

Thorne, Kip S. 1994. *Black Holes and Time Warps: Einstein's Outrageous Legacy.* New York: W. W. Norton and Company.
Thorne, Kip S. 2014. *The Science of Interstellar.* New York: W. W. Norton and Company.

6
Plasticity

In Chapter 4 of his *Ideology and Rationality in the History of the Life Sciences*, Georges Canguilhem notes the etymology of the term *cybernetics*, 'first coined by Ampére in 1834 to denote the science of government,' and which 'lay dormant for more than a century, awaiting the theory that would provide it with a formal concept and enable it to transcend the limitations of its etymology' (Canguilhem 1988, p. 82). Canguilhem goes on to discuss its links to the related term of *regulation*, used to indicate the control of certain organic functions over other functions in the living body. He notes, finally, that any history of our understanding of the biological processes of regulation must incorporate the histories of a range of other sciences (e.g. physics, mathematics, population dynamics, etc.). In the last two chapters, my aim was to examine, from a Bergsonian perspective, the possibilities of thinking the elusive features of processes such as consciousness, and indeed life in general, across the disciplinary boundaries of philosophy, art, and science. To do so entailed a response to Bergson's challenge to bring together the epistemological and ontological exploration of memory; to scrutinize our methods for thinking as a way of uncovering the nature of thought as an evolving, selecting, generalizing, and variously self-alienating and self-liberating dynamic.

W. Khandker, *Process Metaphysics and Mutative Life*,
Palgrave Perspectives on Process Philosophy,
https://doi.org/10.1007/978-3-030-43048-1_6

I want to now set the question of the methods and dynamics of a study of memory (whether human or supra-human) into its broader ontological and ecological context. What does it mean to think 'ecologically' about mind; where 'mind' is meant in the more-than-human senses of the term? I consider this question in four sections. In the first section, I will focus on Hartshorne's essay, 'Minds and Bodies,' in *The Zero Fallacy* in order to review some key features of process thought that set it within a more 'holist' perspective (concerning the cohesiveness of the interactions between communities of processes), with particular reference to Hartshorne's discussions of panpsychism, and the resonance of process with elements of cybernetic theory. In Sect. 2, I will draw on some insights from cybernetics with examples taken from the work of Maturana and Varela on the question of 'behaviour' and 'learning' that cut across human and animal worlds. In Sects. 3 and 4, focusing upon Gregory Bateson's books *Mind and Nature* and *Steps to an Ecology of Mind*, I want to return us to some of the questions posed within evolutionary theories of transmission and transmutation, with which I opened this book, and the ever more urgent problems of the destruction of the environment, asking: What is our place, and what is our future, in our overwhelmingly more-than-human world?

1 The Zero Fallacy

To indicate Hartshorne's position on the relation between mind and body (or mind and matter), a simple yet instructive way to think about it is in terms of what is referred to as 'the zero fallacy:'

> With properties of which there can be varying degrees, the zero degree, or total absence, is knowable empirically only if there is a *known least quantum*, or finite minimum, of the property. Planck's constant is an example; it excludes complete continuity in changes by setting an absolute finite minimum. Thus light intensity may be reduced to *one photon*; less than that is simply no light. I hold that metaphysicians should have anticipated this. Absolute continuity of change, nature 'making no leaps,' never was

> or could be an observed fact, for to observe it one would have to be God, with absolutely clear and distinct perceptions. There are also reasons why God should not be supposed to observe continuous change. A zero of elephants is observable because there is a finite minimum of what can properly be called an elephant. (Hartshorne 1997, p. 166)

As Hartshorne goes on to note, when we consider the difference between mind and matter, or the possibility of their relation, there is, at least in psychology, he contends, an unwillingness to define the 'minimum of mind' (Hartshorne 1997, pp. 166–167). What is the lowest form or level of mind? Before even broaching this question, we first encounter a range of assumptions and conceptual obstacles, such as the epistemological problem of our ability to know other minds (other human minds, let alone the possibility of more-than-human ones); the anthropocentric conceit, if you will, that mind is a distinctively human thing; and the persistence of the idea of the Great Chain of Being in which all forms of life exist in an ascending hierarchy of perfection, from inanimate and mindless matter up to human souls and God. Alternatives to this history of ideas and prejudices can be found in certain panpsychist theories that have themselves persisted over the same period of millennia: we might look to Presocratic thought, Eastern philosophies, and the early modern period in Western thought, in particular Spinoza and Leibniz. It is to this alternative strand of panpsychist thought that many process philosophers appeal and respond. Both Whitehead and Hartshorne cite the Leibnizian monadology as a basis for their thought, in part as an assertion against Kant and the post-Kantian tradition in European philosophy. Process thought entails a rejection of the primacy of subjectivity, not simply in favour of the converse position of the empiricist primacy of objects, but also in the spirit of reformulating our very understanding of mind, intentionality, and agency. Whatever a mind is, it exists always in relation to the things it perceives and conceives. Without this original relation, what we call 'mind' and what we call 'matter' are simply abstractions. Thus, we have the problem of defining the minimum of mind: 'Is it in the lowest animals,' Hartshorne asks, 'or the single-celled organisms, plant and animal, or where? And what physicists seriously defend the notion of mind*less* matter?'[1] (Hartshorne 1997, pp. 166–167).

As I have already mentioned, Hartshorne's approach to this problem ultimately follows the panpsychist tradition, challenging, as he sees it, the historical inadequacy of philosophical attempts to talk about the nature of mind. He notes that 'After three thousand years of trying to find something simply *other-than-mind to talk about* why not try to *talk adequately about mind* in its inexhaustibly various possible forms?' (Hartshorne 1997, p. 150). When it comes to the problem of mind-body dualism, this acknowledgement of the 'inexhaustibly various possible forms' of mind casts light upon the self-inflicted difficulties associated with fixing a distinction between mind and matter before one has even begun to enquire into their possible relation. It is not that one must either simply opt for the primacy of mind, or the primacy of matter, nor, on another level, that one should be locked in the impasse of an insoluble dualism, or fail to provide an account of anything except in the vaguest terms under the auspices of an all-encompassing monism (though the latter might be a charge levelled at process philosophy). Rather, with Hartshorne, we can appeal to a 'process monism' and its implications.

To underline, Hartshorne's proposal is to think of matter and mind as being, in a very real sense, alike, and to think that 'mind in general' exists in a vast number of forms with varying complexity. To support this view, Hartshorne cites the work of key contributors to the 'Modern Synthesis,'[2] and amongst these biologists, Hartshorne favours in particular the population geneticist, Sewall Wright, for his apparent 'psychicalism', noting that the term 'psychicalist' is preferable to 'panpsychism' since the latter 'tends to suggest that anything you please, say a chair, is a sentient individual, which is nonsense' (Hartshorne 1997, p. 134). Yet, as nonsensical as a sentient chair may be, the suggestion is that the difference

---

[1]To repeat Hartshorne's worries about thinking the zero degree is not to ignore other philosophical attempts to contemplate absence (most notably, Sartre's lengthy engagement with this topic in his *Being and Nothingness*). Setting aside any attempt to answer the problem definitively, I will return to the significance of the zero degree when I turn to Bateson's speculations in the final sections of this chapter.

[2]His references include Julian Huxley and Bernhard Rensch; the latter even wrote a work entitled *Biophilosophy*, though the resulting effort is far less convincing than the endeavour promised by the appealing title.

between non-sentient things, such as chairs, and sentient ones, such as human beings, is really only one of complexity of composition: 'Feeling can be everywhere even though many things do not feel, somewhat as vibration can be everywhere even though chairs do not vibrate (only their microconstituents do)' (Hartshorne 1997, p. 134). The ubiquity of 'feeling' is one echoed in Whitehead's theory of prehensions, according to which the minimal unit of activity is the 'taking account of' one entity by another entity, from the moment of a perception of the greenness of grass, up to the collision of an asteroid with a planet. Systems of prehensions (or 'organisms,' as Whitehead calls them), depending on the perspective taken, might describe a conversation between two human beings, or the interactions between a cat and a human being, or the action of a hurricane when it hits a populated city.

Hartshorne's psychicalism, as he names it, is not he claims simply a derivation of our later age's improved knowledge of nature in increasingly microscopic detail, though we have certainly seen a movement towards mechanistic explanations of natural processes reinforced by the findings of physical enquiry. Rather, Hartshorne attributes his psychicalism to 'phenomenological observations of sensations as a special class of feelings' (Hartshorne 1997, p. 165). The idea of 'mindless matter', he continues, is not the object as given to us, but rather an abstraction to which we have ascribed primacy in our descriptions of nature. This deadening of matter might be called, in Bergsonian terms, an intellectual habit. For an alternative to this view of matter, including living matter, Hartshorne summarizes four 'True Leibnizian Principles' that have had enduring value since the early modern period, despite their ultimate obfuscation in Leibniz's own formulation of pre-established harmony between 'active singulars.' He sets out the four Principles as follows:

a. Nature consists of active singulars;
b. What seem to our perception s mere masses of inert, continuous stuff are really composites or assemblages of active singulars too insignificant, taken one by one, to register distinctly on our senses;
c. The only singular activity we can directly *and distinctly* experience as such is our own subjectivity or experiencing;

d. Consequently, our sole hope of understanding nature is to conceive it as a vast society of active singulars, each of which has some analogy with ourselves as sentient individuals, to the extent at least of possessing some form, however different from our own human form, of feeling or experiencing (Hartshorne 1997, pp. 134–135).

Hartshorne's alternative to the pre-established harmony between active singulars is to think of them as sets of relations conceived as 'a creative becoming, where "creative" means (with Bergson, Peirce, Whitehead, and others) that it determines what is undetermined, and thus *adds to the definiteness* of reality' (Hartshorne 1997, p. 135). For this process of determination, he appeals to Whitehead's sense of the term 'decision' which entails the 'cutting off' of possibilities. In the next section I want to explore further this use of the term 'decision' and also of 'behaviour' in the writings of Humberto Maturana and Francisco Varela as an elaboration of some of the problems that Hartshorne has set up above. The resonance between the process and cybernetic terminology will serve to illustrate the movement from a teleological to a 'stochastic' description of evolutionary development to be discussed in Sect. 3.

2 Decision and Behaviour

Maturana and Varela, in *The Tree of Knowledge*, compare plant 'behaviour' (e.g. structural changes from land to aquatic form, in response to changes in environment) with the 'behaviour' of an amoeba, an animal with the most rudimentary of organic structures. The term 'is proper to any unity seen in an environment where the unity specifies a realm of perturbations and maintains its organization owing to the changes of state that these perturbations trigger in it'[3] (Maturana and Varela 1992, p. 142). Maturana and Varela argue that whilst a plant's movements tend to be described as 'developmental' and an amoeba's movements as 'behavioural,' both, in fact, can be legitimately described

[3] Cf. Henri Bergson's discussions, in texts such as his essay 'Life and Consciousness,' in Bergson (1975), of the example of the amoeba to illustrate the lowest level of conscious activity in animal life.

as forms of 'behaviour' (Maturana and Varela 1992, p. 144). Their analysis progresses to their stance on the nature of learning, in which they dispute the idea that learning and memory are processes of 'taking in' from the environment, this latter idea indicating a human prejudice to think biological phenomena within the limits of human capacities such as the capacity for language.

What happens in perception, then, if it is not a 'taking in'? Bear in mind also, that 'taking in' is not meant in the same way as the Whiteheadian definition of prehension as a 'taking account of.' The difference between the two is significant, and I will come back to this point shortly. For Maturana and Varela, perception is a movement of reciprocal exchange rather than a linear process. They describe, evocatively, the process of projection of an image upon the retina not as the transfer of a well-defined object onto a separate, and equally well-defined, surface. Instead, the process is more akin to the addition of a new voice to a cacophony of voices in a 'hectic family discussion (relations of activity among all incoming convergent connections) in which the consensus of actions reached will not depend on what any particular member of the family says' (Maturana and Varela 1992, p. 163). A key implication of this physical process is that the interface that exists between observer and observed is not a one-to-one relation. It is, instead, a convergence of connections in the nervous system (= the observer) with a multiplicity of sensory data which we could, for the sake of description, say 'occurs' at the retina, or that originates in the environment. Writing a description of this complex process forces the demarcation of subject or mind from world, but the difficulty of making such distinctions is interesting in itself. To continue with Maturana and Varela, in these terms the nervous system can be said to 'expand' behaviour rather than to 'invent' or engender it:

> the nervous system emerges in the phylogenetic history of living beings like a network of special cells (neurons), which is embedded in the organism in such a way that it couples points in the motor surfaces. Thus, with a network of neurons coming between this coupling, the field of possible sensorimotor correlations of organism is increased and the realm of behaviour is expanded. (Maturana and Varela 1992, p. 163)

It is in the relation or 'coupling' of the organism with its environment that behaviour emerges and develops in the manner of 'mind.' Furthermore, it is the characteristic of all known nervous systems to display some degree of *plasticity*, something that we might ordinarily or habitually attribute to higher organisms, as we often name ourselves (Maturana and Varela 1992, p. 168). The nervous system is then

> a unity defined by its internal relations in which interactions come into play only by modulating its structural dynamics, i.e., as a unity with operational closure. In other words, the nervous system does not 'pick up information' from the environment, as we often hear. On the contrary, it brings forth a world by specifying what patterns of the environment are perturbations and what changes trigger them in the organism. (Maturana and Varela 1992, p. 169)

On this view, it is necessary both to reject the familiar image of the brain as an information-processing device and to formulate a different description of the process of learning. We might, for example, think of learning as the adaptation of an organism to a change in its environment, thereby reinforcing the dualistic subject-world relation. The operation of the nervous system, however, points to a more complex, decentred, relation in which changes occur that only we, after the fact, reify in certain actions pictured and performed by the organism on the one hand, and in certain environmental changes on the other.

To exemplify the common semantic descriptions of behaviour in contrast to how they would read the operation of the organism, Maturana and Varela describe the process of a cat signalling to its human companion within the context of an ordinary household setting. The human possesses a piano. For the cat, this is a means to express her desire to go outside, and she treads across the keys of the piano, which triggers the human to open the door. In response to this mode of description, Maturana and Varela object that we know, in fact, that 'the interactions between them occurred only as a mutual triggering of changes of state as determined by their respective structures' (Maturana and Varela 1992, p. 206). To speak in terms of what the cat wants, and how the human complies is far too macroscopic in focus. Instead, we should think of the regular interactions between organisms as the generation of a social coupling in which 'they are reciprocally involved in attaining their respective poieses. Behaviors that take place in these domains of social coupling, as we said, are communicative and they can be inborn or acquired' (Maturana and Varela 1992, p. 206). As we can see, for Maturana and Varela, the terms of our description must themselves change, resulting, for example, in an account where there is in the nervous system 'only an ongoing structural drift that follows the course in which, at each instant, the structural coupling (adaptation) of the organism to its medium of interaction is conserved' (Maturana and Varela 1992, pp. 170–171). Thus, this 'cybernetic' approach disputes the idea of learning and memory as a process of 'taking in' from the environment, since 'to describe learning as an internalization of the environment confuses things by suggesting that in the structural dynamics of the nervous system there are phenomena that exist only in the descriptive realms of some organisms, like ourselves, capable of language' (Maturana and Varela 1992, p. 172). This challenges our tendency to describe the cat-human relation in terms of two distinct species, who normally operate in quite separate semantic domains, somehow mysteriously bridging their species gap and achieving the impossible, 'communication.'

If we turn to Gregory Bateson at this point, we can see how he elaborates upon this enquiry into non-anthropocentric ideas of communication and mind. Bateson (in his *Mind and Nature: A Necessary Unity*) reiterates Hartshorne's concerns about the problematic scientific resistance to finding the 'minimum of mind,' but he shows that this has been largely due to a category mistake. Mind should be understood as *pattern* rather than as a thing or substance. To illustrate this mistake, Bateson highlights a fallacy that has persisted in the histories of philosophy of science connecting response and change to 'energy.' That is, where a change occurs, a metaphor of energy is often invoked as the driver or source of the change or movement that follows. Descriptively, this is unhelpful in cases where there actually appears to be a significant difference between zero and one: for example, a lack or deficiency of nutrients in the environment can itself provoke a response, a change of position, in an amoeba.

For Bateson, pattern and form are more informative for thinking about the nature of mind (in all things; at all levels). He illustrates this in the case of evolution by natural selection, just one level of description invoked in cybernetic writings, to show that cybernetics provides a *negative* form of explanation. In it, 'we consider what alternative possibilities could conceivably have occurred and then ask why many of the alternatives were not followed, so that the particular event was one of those few which could, in fact, occur' (Bateson 1987, p. 405). As he continues, 'The cybernetic method of negative explanation raises the question: Is there a difference between "being right" and "not being wrong"? Should we say of the rat in a maze that he has "learned the right path" or should we say only that he has learned "to avoid the wrong paths"?' (Bateson 1987, p. 411). What Bateson is hinting towards is how we might introduce the ideas of intention or direction or mind into our concept of nature. It would then be possible to think of a level at which positive choices are made, rather than choices to discard unsuitable pathways. This suggestion recalls the 'anti-Darwinian' schools of thought that attempted to invoke some degree of directionality into evolutionary development, such as Lamarck's theory of the transmission of acquired characteristics, or the proponents of orthogenesis who sought a law of evolutionary change at the genetic level rather than abandoning such an important mechanism to chance in the manner of the theory of natural selection. As unpopular a thing as intention is in scientific circles, Bateson's interest is in acknowledging something that philosophers (and theologians) might call, in less subtle terms, design. In Bateson's less loaded terminology, he suggests the following:

> If, in the communicational and organizational processes of biological evolution, there be something like *levels*—items, patterns, and possibly patterns of patterns—then it is logically possible for the evolutionary system to make something like positive choices. Such levels and patterning might conceivably be in or among genes or elsewhere. (Bateson 1987, p. 411)

Within this account is an argument against the idea that human language is an evolutionary step ahead of other animal forms of communication (kinesics and paralanguage). On the contrary, the latter have in fact evolved in complex ways alongside verbal language.[4]

The 'meaning' that is the focus in the study of language, is replaced at the level of biological systems by the idea of 'Redundancy.' Patterning (in the sense used by engineers and mathematicians) in message material 'is said to contain "redundancy" if, when the sequence is received with some items missing, the receiver is able to guess at the missing items with better than random success' (Bateson 1987, p. 419). Meaning can be inferred from a sequence of units, even when we are not given the whole picture, for example, numbers in a sequence; and further, 'camouflage (the opposite of communication) is achieved (1) by reducing the signal/noise ratio, (2) by breaking up the patterns and regularities in the signal, or (3) by introducing similar patterns in to the noise' (Bateson 1987, p. 420). This account of the existence of redundancy in biological systems compares readily with Bergson's descriptions of the *élan vital* as encoded in the bodies of individual organisms over long periods of evolutionary development, with the invisibility of the underlying processes now formulated in terms of the operation of information systems. Bateson refers to this process as phylogenetic learning:

> There is redundancy in the system, organism-*plus*-environment, such that from the morphology and behaviour of the organism a human observer can guess with better than random success at the nature of the environment. This 'information' about the environment has become lodged in the organism through a long phylogenetic process, and its coding is of a very special kind. The observer who would learn about the aquatic environment from the shape of a shark must deduce the hydrodynamics from the adaptation which copes with the water. The information contained

[4]As Bateson explains further: 'Both kinesics and paralanguage have been elaborated into complex forms of art, music, ballet, poetry, and the like, and, even in everyday life, the intricacies of human kinesic communication, facial expression, and vocal intonation far exceed anything that any other animal is known to produce. The logician's dream that men should communicate only by unambiguous digital signals has not come true and is not likely to. [...] This burgeoning evolution of kinesics and paralanguage alongside the evolution of verbal language indicates that our iconic communication serves functions totally different from those of language and, indeed, performs functions which verbal language is unsuited to perform' (Bateson 1987, p. 418).

> in the phenotypic shark is implicit in forms which are complementary to characteristics of other parts of the universe, *phenotype plus environment* whose redundancy is increased by the phenotype. (Bateson 1987, p. 422)

I will return to the significance of observing, understanding, and enhancing *human* life in terms of the relation *organism-plus-environment* in the final section of this chapter.

3 Stochasticity

I want to turn, first, to a more detailed examination of the relation between the two systems named 'organism' and 'environment.' We have seen that their relation is not one of mind and matter, active environment and passive organism (nor indeed the reverse: organism acting on world), nor simply stimulus and response. The interactions of many complex systems require different modes of description. One such alternative comes, for example, as Hartshorne notes in relation to the science of ethology, in the move from deterministic explanation to *stochastic* explanation, a change that implies '*not* the abandonment of the causal principle, but its *reformulation*' (Hartshorne 1997, p. 135). Bateson sets out some parameters for a discussion of stochasticity in the relation between organism and environment in evolutionary processes. The survival of any organism is conditional upon a complementarity of two main types of change: firstly, an inner conservatism of the body that can only tolerate a limited amount of disruption to its integrity[5]; secondly, alterations in the external environment may require 'a sacrifice of conservatism' for the organism to prevail (Bateson 1979, p. 144). Bateson's subsequent exemplification of these two processes in the contrast between the presence of homologies, a result of conservatism, and the phenomenon of adaptation, which speaks to a responsiveness to change, revisits the problem of recapitulation (Haeckel, D'Arcy Thompson), but the scope of

[5]Cf. Niels Bohr's discussion of the invasiveness of physical investigations that can destroy the thing it seeks to discover: the 'vitality' of a living organism.

Bateson's enquiry is not to reinstitute developmental hierarchies (phylogenetic trees). Rather, Bateson favours a 'Platonic study' of the parallels between creative thought and biological evolution (echoing Bergson's comparison between 'life and consciousness'), in which we ask the following questions:

> Is *this* way of looking at the phenomena somehow represented or paralleled within the organizational system of the phenomena themselves? Do any of the genetic messages and static signs that determine the phenotype have the sort of syntax (for lack of a better word) which would divide 'typological' from 'synthetic' thinking? Can we recognize, among the very messages which create and shape the animal forms, some messages more typological and some more synthetic? (Bateson 1979, p. 171)

Compare now our two stochastic subsystems within the larger system of evolution. Bateson suggests a complex relationship between subject and world (or organism and environment) operating at a number of levels. First, at the level of the dynamics of internal selection, the organism's body is the stage for a conflict between its default conservatism and its capacity for mutation in the natural processes of DNA replication; it incorporates 'a random [genetic] component and a process of selection working on the products of the random component' (Bateson 1979, p. 176). Second, in the dynamics of external adaptation, the external manifestation (phenotype) of an organism's genetic identity (genotype) is subjected to the pressures of environmental change, and 'the random component is provided by the system of phenotype in interaction with environment' (Bateson 1979, p. 178). The interaction of these two levels is the dynamic of natural selection.

In this complex process, neither the organism nor the environment is by itself the driver of evolution. Answering an objection to the view of natural selection as entirely random, Bateson notes that Darwin made a concession to Lamarck with an acknowledgement that geological time was insufficient to allow for the changes he theorized in natural selection. Yet a solution to the problem of apparent randomness comes through a reconsideration of scales of interaction. It is not that natural selection works on individual genes, or even on individual members of a species, but rather on *populations*:

> Dobzhansky's discovery that the unit of evolution is the population and that the population is a heterogeneous storehouse of genic possibilities greatly reduces the time required by evolutionary theory. The population is able to respond immediately to environmental pressures. The individual organism has the capacity for adaptive somatic change, but it is the population that, by selective mortality, undergoes change that is transmitted to future generations. (Bateson 1979, p. 179)

Internal selection (on genotype) is likely to take longer to manifest external changes (phenotype). Bateson notes the resemblance in cytology across species, and embryological development, as opposed to the great differences between phenotypes, and concludes that it must be more difficult to effect changes or 'deviations' in the embryological stages than in later stages of life (Bateson 1979, p. 180). Furthermore, 'The machinery of change is not simply permissive or simply creative. Rather, there is a continual determinism whereby the changes that can occur are members of a *class* of changes appropriate to that particular machinery. The system of random genetic change filtered by the selective process of internal viability gives to phylogeny the characteristic of pervasive homology' (Bateson 1979, p. 181). It is the gap between the somatic/analogic (solely quantitative changes) and the genetic/digital (variously qualitative or quantitative changes) that ultimately prevents a Lamarckian mechanism of inheritance (Bateson 1979, p. 182).

In a note, at the end of the section on Ecology and Epistemology in *Steps to an Ecology of Mind*, Bateson stresses his disagreement with Samuel Butler, Whitehead, and Teilhard de Chardin 'that it follows from this mental character of the macroscopic world that the single atomies must have mental character or potentiality' (Bateson 1987, p. 472). Rejecting the idea that the character of mind applies *all the way down*, in the manner of Whiteheadian prehensions, Bateson affirms the emergence of mind in the formation of more complex structures. His unwillingness to align himself with Whitehead on this point reveals a hasty and unjustified criticism of the philosophy of organism. Whitehead's system of prehensions seeks precisely to overcome the Leibnizian formulation of monadic relations in favour of a revised conception of mind as pure relation. It is not that mind (particularly human mind) is self-contained enough to serve as starting point in any enquiry into the nature of mental activity. Rather, mental activity occurs, and should be reconceived at the microscopic level as a system of relations of 'taking account of.' These microscopic activities, or prehensions, underpin and compose other systems at a macroscopic level, such as individual human beings, or any other individual member of a species, or a whole society of like individuals.

On Bateson's part, if there is an affinity between the processes of evolution by natural selection and the processes of human thought, as Bergson in particular propounds, it is based on their common 'stochastic' nature. Bergson helps to problematize the hierarchy of human intelligence above other species, by making human mind and creativity an instance or expression of a broader and more primal movement of an *élan vital*. Human intelligence is not an exception that sets us apart from nature, but rather one pathway amongst many taken by life in the course of its development across divergent lines of evolution. Whitehead goes on to problematize the very idea of mind by locating perception and feeling not in the subject, but in the myriad moments in which objects relate to one another, at all levels. What counts as a subject can be drawn with a multiplicity of borders, according to the level of analysis. Bateson's worry that this theory of prehensions places mental activity everywhere does not appreciate the implications of Whitehead's scheme for our understanding of the human mind.

Bateson's own approach to rethinking mind, and the contribution his view makes to a dissolution of the exceptionalism of human intelligence, nevertheless highlights a parallel between the creative process of evolution and the creative process of human thought. Commenting upon the two stochastic systems, he explains that what ensures a parallelism between the processes of biological evolution, on the one hand, and the processes of mind, on the other, is not their creation by a 'Designer' indiscernible yet present somewhere in the 'machinery' of evolution. It is instead their common basis in a certain stochasticity:

> *creative* thought must always contain a random component. The exploratory processes—the endless *trial and error* of mental progress—can achieve the *new* only by embarking upon pathways randomly presented, some of which when tried are somehow selected for something like survival. (Bateson 1979, p. 182)

What, then, does it mean to think 'ecologically' about human and non-human animal minds? In the preceding sections I have set out to consider this question in terms of (i) the traditional philosophical problem of mind-body dualism, (ii) the process thought response to this problem in the distribution of 'mind' throughout nature (one itself that is rooted in panpsychism), (iii) some speculations on what this decentralized, anti-anthropocentric view of mind implies for our accounts of behaviour in human and non-human organisms alike, (iv) and the senses in which this model of mind can be said to operate at many levels of generality, thus not simply in the creative activities of individual minds, but also at the level of the supra-individual creative process that we call evolution by natural selection. I want to conclude this chapter with a comparison of Bateson's 'ecology of mind' with one final recourse to Hartshorne on the nature of subjectivity.

4 Psychicalism and the Ecology of Mind

At this stage, let us be reminded that we are no longer dealing with a view of conscious subjectivity in the sense of an individual human subject, but instead with a more diffused and 'nonanthropomorphic' sense of subjectivity which describes 'anything that can be said to be aware of (know or feel or intuit) anything' (Hartshorne 1997, p. 138). By 'subject' we refer now to the 'subjective pole' of a subject-object relation, in which 'the state, not the substance, the experience (in its aspect of awareness *of* something) not the ego, is the subject' (Hartshorne 1997, p. 138). In this scheme, objects do not depend on subjects for their determination. Subjects, whether human or non-human, are neither simply 'constitutive' (as Kant would have it) nor 'world-forming' (as Heidegger would have it). Instead, subjects are always determined by the objects of which they are aware. Subjectivity is made. It becomes. All entities are, or are capable of being objects for other subjects; all subjects are 'images' in the Bergsonian sense. Finally, all 'concrete' entities are subjects. Hartshorne refers to this as the complementarity of the theses of Objective Independence, Subjective Dependence, Universal Objectivity, and Universal Subjectivity (or panpsychism) (Hartshorne 1997, pp. 138–139).

The challenge for us (*Homo sapiens*) is to overcome our 'narrowness of experience,' and to appreciate that subjectivity, mind, or consciousness operates beyond the bounds of a definition of consciousness that is derived primarily from an inspection of our own minds:

> The illusion to be overcome is not, 'all reality is experience, feeling, subjectivity, value,' for this is no illusion, but the veritable *egocentric illusion,* 'experience, feeling, subjectivity, values, are solely or chiefly found in my self, or my kind of self.'[6] (Hartshorne 1997, p. 149)

Arthur Lovejoy's book *The Great Chain of Being* traces the development and persistence of the ancient Greek idea of the organization of nature in terms of degrees of perfection, an idea that saw human life placed above and as a standard against which to measure all other forms of life. The persistence of the idea of the great chain of being across millennia makes, according to Bateson, Lamarck's claims in texts such as the *Physiologie Zoologique* (1809) all the more extraordinary for his call to reverse the taxonomic hierarchy showing that life 'starts with the infusoria and that there were changes leading up to man' (Bateson 1987, p. 433). Bateson sees in Lamarck a number of deceptively modern ideas, such as the claim 'that you cannot attribute to any creature psychological capacities for which it has no organs; that mental processes must always have physical representation; and that the complexity of the nervous system is related to the complexity of mind' (Bateson 1987, p. 434). The nature of this complexity was also indicated by Alfred Russel Wallace, who compared

[6] Hartshorne defends this position as a 'true physicalism.' It is not simply that he is a 'spiritualist' or idealist, but that what we understand physicality and spatial position to be require reformulation: 'If physical means spatial then mentalism or psychicalism is physicalism, for space is how sentient beings have neighbors (Peirce) with whom they react, and their basic operations (Whitehead) are prehensions, feelings of (others') feelings' (Hartshorne 1997, p. 150).

natural selection, in an essay sent to Darwin, to 'the steam engine, which checks and corrects any irregularities almost before they become evident' (Bateson 1987, p. 434). We return here to the definition of cybernetics, and we find in Wallace's essentially cybernetic explanation a means of understanding the complexities of life and mind:

> The steam engine with a governor is simply a circular train of causal events, with somewhere a link in that chain such that the more of something, the less of the next thing in the circuit. The *wider* the balls of the governor diverge, the *less* the fuel supply. If causal chains with that general characteristic are provided with energy, the result will be… a self-corrective system. (Bateson 1987, p. 435)

Now that we have a model for thinking about the nature of mind, Bateson's 'ecological' approach underlines how each system works in harmony with the systems that surround it, and ultimately with the larger ecosystem in which it operates. The stronger claim here, echoed in Lovelock's Gaia hypothesis, is that it should be the aim of each component of a complex system to take account of the 'whole systemic structure.' However, this balance is not always achieved. The environmental harms perpetrated by the human species, are an indicator of this failure to operate 'mindfully' or within the wider complex system. Similarly, Henri Bergson notes that human beings possess a utility-driven form of perception, and that this mode of selecting from our immediate environment only those aspects that are useful to us (that will aid us in completing our present task), is simply the human manifestation of a characteristic of all animate life: the effort to manipulate matter. Problems arise when we mistake this 'carving up' of the world according to its usefulness for the inherent constitution of things in themselves. Yet human consciousness is so constituted, that it cannot help but follow this ever-narrowing path.[7]

[7]The following account from Bateson could just as easily be lifted from one of Bergson's texts: 'Consciousness operates in the same way as medicine in its sampling of the events and processes of the body and of what goes on in the total mind. It is organized in terms of purpose. It is a short-cut device to enable you to get quickly at what you want; not to act with maximum wisdom in order to live, but to follow the shortest logical or causal path to get what you next want' (Bateson 1987, p. 439). Bergson's biophilosophy is by no means a popular theory

Bateson provides two examples of this imperfect adaptation of human activity with its wider environment. The first example concerning 'ecosystem imbalance' illustrates in structural terms some potential problems that might confront the adaptation of human societies with individuals belonging to adjacent ecosystems. Using analogies drawn from Lewis Carroll's *Alice in Wonderland* Bateson explains that:

> By imperfect coupling of biological systems in the famous game of croquet, however, Carroll creates a meta-random game. Alice is coupled with a flamingo, and the 'ball' is a hedgehog. [...] Alice's difficulty arises from the fact that she does not 'understand' the flamingo, *i.e.,* she does not have systemic information about the 'system' which confronts her. Similarly, the flamingo does not understand Alice. They are at 'cross-purposes'. The problem of coupling man through consciousness with his biological environment is comparable. (Bateson 1987, p. 449)

It is debatable whether Bateson is participating in human exceptionalism when he describes human action on the environment as 'imperfect coupling of biological systems,' which affirms anthropogenic changes as highly distinctive compared to other species' impacts (in his albeit critical perspective on the human relationship to the rest of nature). Rhetorically at least, this is a compelling evolutionary account of certain human activities, such as the cultivation of its own symbionts and parasites in the effort to maintain the dominance of its own species, and 'these, likewise, become single-species ecosystems: fields of corn, cultures of bacteria, batteries of fowls, colonies of laboratory rats, and the like' (Bateson 1987, p. 451). Indeed, the more recent shift towards the promotion of biodiversity as the primary goal of conservation, and the hard-won recognition of human activity as driver of climate change, suggests that there is a great degree of 'imperfection' in our synchronization with our environment. As part of this misalignment, Bateson also takes to task the advance of medical knowledge down certain routes, citing it as a strong example of a purpose-driven science (i.e., the search for cures) which acts counter to 'wisdom' about biological systems:

---

in contemporary philosophy of biology, but it is one that nevertheless proves its potency and relevance in dialogue with the much later cybernetic stance presented by Bateson.

> Medicine ends up, therefore, as a total science, whose structure is essentially that of a bag of tricks. Within this science there is extraordinarily little knowledge of [...] the body as a systemically cybernetically organized self-corrective system. Its internal interdependencies are minimally understood. What has happened is that purpose has determined what will come under the inspection or consciousness of medical science. (Bateson 1987, p. 439)

The 'epistemological distortions' of a utility-driven science can serve as a barrier to our adaptation to deteriorating environmental conditions.

Hartshorne warned against the 'egocentric illusion' that mind was a characteristic of human life and not of other forms; to this we might add Bergson's diagnosis of our tendency towards dogmatism in the reification of our intellectual habits at the expense of immediate, sensory experience; and with Bateson we might caution against ignoring 'systemic wisdom.' Yet, we return to the problems raised by the Gaia Hypothesis: If cybernetic systems are, like the engine governor, self-corrective systems, is it not the case that human activity, no matter how destructive, will ultimately find itself subject to the whole system's corrective measures? The answer is, of course, 'yes.' Biological systems, if disturbed, will 'punish' those component systems that fail to behave in harmony with it (Bateson 1987, p. 440). This, for Bateson, is a statement about the predicament of the human species in its current state of environmental crisis, demonstrating that 'a further correction of the unit is necessary. The flexible environment must also be included along with the flexible organism because [...] the organism which destroys its environment destroys itself' (Bateson 1987, p. 457).

Regaining this status as flexible component within the autopoietic system requires a radical shift in perspective on our fellow components (the complex of other species and our geological environment). Bateson cites here the Pythagorean commitment to an understanding of the world in terms of numerically repeating patterns, as opposed to its grounding in some sort of substance, that lays the groundwork for a process understanding of reality, but one that is grounded in formal or mathematical principles. Pythagoras was regarded more as a mystic than a philosopher, and we might certainly agree on inspection of aspects of his thought such as the doctrine of the transmigration of the soul, or metempsychosis, that paints an image of the universe in which each new death and rebirth into life sees the soul become a different creature.[8] However, in its anticipation of the Platonic idea that all coming to know is really a reminiscence (*anamnesis*) of what is forgotten from a previous life, a Pythagorean grounding of reality in pattern or form, and the repetition of it in time across generations, at its simplest rehearses what is later formulated as physical law (Collingwood 2014, p. 53); as the existence of innate ideas; the Kantian model of a priori structures of our cognitive faculties; or innate psychological tendencies to recognize certain forms.[9]

Bateson's appeal to Pythagoras as a way of articulating the extent of the metaphysical shift required in order to enact a biocentric outlook, chimes with D'Arcy Thompson's efforts to explain the extent of physical constraint on biological processes. The aim is not to reduce the biological wholly to the physical, but to locate as many alternative commonalities between human and more-than-human life that might show us to be set on complementary, and not conflicting, paths. Ultimately, 'it makes no difference what the world is made of, and that what we have to study is the patterns and changes of pattern which this primitive matter, whatever it may be, adopts and undergoes' (Collingwood 2014, p. 53). Taking the idea of pattern forward with a more modern inflection incorporates a view of the world in terms of a relation, and sometimes

[8]See Barnes (2001). As Porphyry, in his *Life of Pythagoras*, tells us: 'But it became very well known to everyone that Pythagoras said, first, that the soul is immortal; then, that it changes into other kinds of animals; and further, that at certain periods whatever has happened happens again, there being nothing absolutely new' (Barnes 2001, p. 33).

[9]In his *Commentary on the* Physics, Simplicius includes a summary of Pythagorean thought: 'One might wonder whether or not the same time recurs, as some say it does. Now we call things "the same" in different ways: things the same in kind plainly recur – e.g. summer and winter and the other seasons and periods; again, motions recur the same in kind – for the sun completes the solstices and the equinoxes and the other movements. But if we are to believe the Pythagoreans and hold that things the same *in number* recur – *that you will be sitting here and I shall be talking to you, holding this stick, and so on for everything else* – then it is plausible that the same *time* too recurs' (Barnes 2001, p. 35). See also R. G. Collingwood (2014): '[Pythagoras] showed that the qualitative differences between one musical note and another depend not on the material of which the strings producing these notes are made but solely on their rates of vibration: that is to say, on the way in which any given string successively, in a regular rhythm, takes up a determinate series of geometrical shapes. Alter the tempo of this rhythm and you alter the note; produce the same rhythm in two different strings, and you make them both yield the same note' (Collingwood 2014, p. 52).

conflict, between two key aspects of ecological thought: 'the face which is called bioenergetics—the economics of energy and materials within a coral reef, a redwood forest, or a city—and, second, an economics of information, of entropy, of negentropy, etc.' (Bateson 1987, p. 466). The former concerns boundaries at cell membrane, skin, or individual level. For example, in Darwinism, the unit of selection is the breeding individual, the family line, or the sub-species (or society), but if it is the aim of any evolutionary progression to serve only that individual unit, then, as we saw earlier, it will end in a destroyed environment, thereby destroying itself (Bateson 1987, p. 457). On the other hand, thinking in terms of the economics or flows of information entails a concern with pathways and probabilities, such that what we mean by 'mind' is therefore the total circuit, or cybernetic system, moving between differences in internal and external relations:

> The individual mind is immanent but not only in the body. It is immanent also in pathways and messages outside the body; and there is a larger Mind of which the individual mind is only a subsystem. This larger Mind is comparable to God and is perhaps what some people mean by 'God', but it is still immanent in the total interconnected social system and planetary ecology. (Bateson 1987, p. 467)

In this final comment from Bateson on individual mind as manifesting in microcosm what the Earth's ecosystem (or Gaia) is in macrocosm we also return to Whitehead's philosophy of organism. The value of our starting point in Whitehead's vision of communities of prehensions that compose 'organisms' at varying scales of being was to form the basis for an ecological acknowledgement of activity across all orders of life, and across all non-living systems. The difficulties set out in the earlier chapter, 'Transmutation,' concerning the measurement of fluctuating populations, and the designation of certain species as despotic and invasive, on the one hand, and vulnerable or endangered on the other, are confronted here with a challenge to *Homo sapiens* to rethink its own impact and to attune itself better to its environment. We can respond to this challenge first in the manner of Bergson's reconfiguration of mind or memory as a process that functions beyond the confines of cerebral matter. Thinking about mind, and processes more generally, 'obliquely' helps us to orient ourselves differently both to our own capacities and to those of other species whose conscious functions do not manifest in relation to that particular configuration of the nervous system that defines, at least, taxonomically, our own species. In fact, integral to this re-orientation towards human consciousness is precisely a renewed perspective on the distribution and diffusion of different intensities of consciousness that Bergson speaks of in *Creative Evolution.* As Hartshorne reminds us, our inability to identify 'the minimum of mind' is a particular feature of debates on the distinction between mind and matter that opens up the possibility of conceiving some form of rudimentary activity as the starting point for any theory of mind. Having taken this step, Hartshorne's recourse to 'psychicalism,' and to radically non-subjective accounts of activity such as Leibniz's monadology, retraces the path followed by a long tradition of philosophers towards more 'diffused' concepts of mind. The strictly chemical and biological concept of Gaia is explicable in more philosophical terms that can be united with thinkers such as Whitehead, Bergson, and Hartshorne, through Bateson's explicit appeals to the history of philosophy for a broader range of concepts that might help us to achieve a greater harmony with our environment.

References

Barnes, Jonathan. 2001. *Early Greek Philosophy.* London: Penguin.

Bateson, Gregory. 1979. *Mind and Nature: A Necessary Unity.* London: Wildwood House.

Bateson, Gregory. 1987. *Steps to an Ecology of Mind.* Northvale, NJ and London: Jason Aronson Inc.

Bergson, Henri. 1975. *Mind-Energy: Lectures and Essays,* trans. H. Wildon Carr. Westport, CT: Greenwood Press.

Bergson, Henri. 1998. *Creative Evolution,* trans. Arthur Mitchell. New York: Dover.

Canguilhem, Georges. 1988. *Ideology and Rationality in the Life Sciences,* trans. Arthur Goldhammer. Cambridge, MA: The MIT Press, 1988.

Collingwood, Robin George. 2014. *The Idea of Nature*. Mansfield Centre, CT: Martino Publishing.

Hartshorne, Charles. 1997. *The Zero Fallacy and Other Essays in Neoclassical Philosophy*. Chicago and La Salle: Open Court.

Maturana, Humberto R., and Francisco J. Varela. 1992. *The Tree of Knowledge: The Biological Roots of Human Understanding*. Boston and London: Shambhala.

7

Extinction

Let us return to Charles Darwin's derogatory statements about Australian marsupials in *The Origin of Species*, 'feebly representing' the division of mammalia into carnivores, ruminants, and rodents in the northern hemisphere (Darwin 1996, pp. 95–96). Since these animals are older, they are therefore deemed to be less evolved than their European counterparts, and with it more susceptible, even deserving, of decline and extinction. However, twentieth-century developments in global conservation movements (granting that they constitute a machine with many moving, and not always harmonious, parts) have embedded in popular discourse an understanding of extinction as a harm to be avoided, be it due to anthropogenic or non-anthropogenic causes. How we have moved from the first characterization of 'natural extinction' to the second view of extinction as harm, and indeed, how the first attitude still to some extent 'contaminates' the second reveals some unresolved conceptual problems underpinning modern conservation practices.

One key problem is the persistence of hierarchical thinking, as illustrated by Stephen Jay Gould in his musings over the now often overlooked career of American naturalist, Nathaniel Shaler, in his own time already eclipsed somewhat by his association with Louis Agassiz, the

W. Khandker, *Process Metaphysics and Mutative Life*,
Palgrave Perspectives on Process Philosophy,
https://doi.org/10.1007/978-3-030-43048-1_7

influential and fervent opponent to Darwin's theory of evolution by natural selection. Shaler, like Agassiz, advocated for a taxonomy of species in line with some concept of a natural order, 'as thoughts of a Creator, and taxonomic relationships as the interconnections within His mind' (Gould 1991, p. 315). Shaler's eventual embrace of some form of evolutionism would be Neo-Lamarckian in spirit, accentuating the link between adaptive success and individual effort, which would ultimately take on racist overtones in the affirmation of hierarchies of human races as species-level differences (Gould 1991, p. 316). An intervention into this line of thinking comes from another student of Agassiz, the philosopher William James. In a rebuttal to Shaler's argument that human evolution is an indicator of God's benevolence, James pours scorn onto the idea that any knowledge of probabilities can be gained from the observation of just one outcome of the process:

> We never know what ends may have been kept from realization, for the dead tell no tales. The surviving witness would in any case, and whatever he were, draw the conclusion that the universe was planned to make him and the like of him succeed, for it actually did so. [...] The world has come but once, the witness is there after the fact and simply approves. (James, cited by Gould 1991, pp. 319–320)[1]

The spectre of human exceptionalism thrives by virtue of an affirmation of the superiority of later evolving species over earlier arrivals.

1 *Thylacinus Cynocephalus*

Consider some of the literature on the specific case of *Thylacinus cynocephalus*, or thylacine, which is generally thought to have become extinct in the 1930s. The thylacine, otherwise variously known as the marsupial tiger or wolf, was a member of the family Thylacinidae, under the order Dasyuromorphia that also encompasses still extant members of the family Dasyuridae (Attard et al. 2011, p. 293). The family of dasyurids

[1]See Gould, 'In a Jumbled Drawer,' in *Bully for Brontosaurus* (1991) on James, Agassiz and Shaler.

includes the endangered Tasmanian devil (*Sarcophilus harrisii*) and the spotted-tailed quoll (*Dasyurus maculatus*), both notable examples of marsupial carnivorism that enable some insight into what might have been the life and habits of the somewhat larger carnivorous thylacine. Debate about the causes of decline of the thylacine continues, but the species is thought to have been widely distributed on mainland Australia, as well as the island habitat of Tasmania, until the arrival of dingoes around 4000 years ago. Tasmania's isolation from that particular incursion of competing carnivores helped maintain a sizeable population of thylacines there until the era of European colonization commencing in the late eighteenth century (Jones and Stoddart 1998, p. 239).

Wemmer (2002) observes that the time spent by thylacines in zoos ought to have yielded significant insights into the species: 'String their lives together and you have more than 100 thylacine years in captivity: ample opportunity for some keen observer to note a few details' (Wemmer 2002, p. 2). Yet very little is known about it. Tellingly, Wemmer's opening remarks reveal something about the persistence of the subordinate status of this animal. This bad analogue of a tiger, overlapping with its superior namesake only for its striped coat and carnivorous diet, 'was actually a "wannabe" wolf.' This strange animal had a confusing anatomy: the front half of its body possessed seemingly canine features mixed with a rear half, including the pouch and very long tail, that indicated a marsupial heritage. Yet it was not a 'true dog' for the rather innocuous fact that it was unable to wag its tail. Its one redeeming feature was that it was 'a remarkable example of convergent evolution, and the largest carnivorous marsupial to survive into the 20th century' (Wemmer 2002, p. 1). The unflattering turns of phrase used to describe this animal are merely reflections of older, discredited attitudes, and a more significant reflection of this can be found in Robert Paddle's book *The Last Tasmanian Tiger* which provides the first comprehensive survey of the natural history of the thylacine, tempered with a certain scepticism towards the authority of scientific descriptions of the species in its last years of existence. As Paddle notes, it is customary to defer to the most recent pieces of scientific literature, in line with the method of review and refinement that defines the cutting edge of scientific theory. However, in the case of extinct species, the most recent pieces of analysis of animals whilst they were still alive would have been undertaken on specimens that were 'likely to be operating under severe psychological stress' (Paddle 2002, p. 9). Decimated populations in the wild and individuals kept in captivity ('inappropriate contexts') are not the best sources of accurate information about a species, and yet, Paddle continues, the tendency to prioritize records from the twentieth-century (the last three decades of the species' existence) persists (Paddle 2002, p. 9). Throughout his study, Paddle's sceptical tone helps to communicate key observations about the thylacine, whilst raising important questions about the practices of biologists, conservationists, and policy-makers both historically and in relation to current and future extinction crises.

What do we know about thylacines? It seems the best insights into the lives of this species come from comparisons with two other groups of animals: still extant members of the related family of dasyurids (devils and quolls), and analogously with placental mammals in the canine and feline families. Note that the Latin rendering of the species and genus, *Thylacinus cynocephalus*, roughly and clumsily translates as 'dog-headed pouched-dog' (Paddle 2002, p. 7), whilst common names and comparisons for the species range from tigers, to wolves and hyenas. The 'tiger-stripes' of the thylacine suggest something about the predator-prey dynamics within which the species evolved. Rear stripes are evolved through a predator-prey relation in which slower prey are singled out by the initial alarming presence signalled by the bold colouration of the predator's body and distinctive markings on the face (orientation of contrasting colours indicate the direction in which the animal is facing and thus its intent). Such a dynamic, in order to have evolved (through mutual advantage), must benefit both predator and prey. The bold stripes benefit prey by alerting faster individuals to the threat, and they in turn benefit the predator by identifying slower animals for capture, and thereby helping them to avoid unnecessary expenditure of energy on futile chases. Observation of placental tigers and other big cats seems to bear out this thesis (Paddle 2002, pp. 42–43). In terms of behaviour, the thylacine's social structure seemed most similar to smaller members of the family Canidae, such as foxes and coyotes, forming 'nuclear families' of a breeding pair and three or four cubs.

However, certain distinguishing features of the thylacine are better analysed in comparison with other marsupials. For example, the thylacine was notable for its tendency to greet strangers (of its own, or other species) with a wide 'threat-yawn,' also observed in Tasmanian devils. The observation of distinctive marsupial characteristics of the species, however, also succumbs to the tendency towards 'placental chauvinism,' as we saw in Charles Darwin's broadly derogatory remarks about Australian fauna. Paddle suggests placental chauvinism was one of three key factors that contributed to the thylacine's extinction, the others being a certain 'colonial inferiority', and blatant racism infecting the relations between the Europeans, the settlers, and the colonized (both human and animal) (Paddle 2002, p. 78).

The assumption that marsupial reproduction is an evolutionary intermediary between egg-laying species and placental species is coupled with assumptions such as the relative rigidity of the thylacine's long (kangaroo-like) tail being a betrayal of primitiveness compared to more 'developed' mammals such as dogs and cats, who can wag or articulate their tales more expressively. Paddle also notes the tendency in twentieth-century reports to portray the thylacine, a diurnal (daytime) hunter, as a silent, nocturnal animal, whereas nineteenth century accounts reported a range of vocalizations[2] (Paddle 2002, p. 172). Whilst the scientific identification of the thylacine's features was decidedly negative, it was cemented in the social and political attitudes towards the species, almost certainly hastening its extinction. Paddle's survey of nineteenth century records of the Van Dieman's Land Company's attempts to establish sheep farming settlements in what is now known as Tasmania suggests that a case of local management failure inadvertently led to the extinction of a species. The introduction of a series of bounties aimed at a number of predatory animals augmented the sense of threat posed to sheep-farming by the thylacine, whilst actual losses of livestock were more likely attributable to the Company's own sheepdogs and dogs of white settlers (Paddle 2002,

[2]Paddle recounts the social and political pressure applied in the drafting of a scientific report published in the 1980s to leave out any mention of vocalisations. Discourse on the purported decline of the species was contaminated by the contention that scientists could not find the thylacine because it was so quiet, and not because it was nearly extinct! (Paddle 2002, pp. 171–172).

pp. 114–115). With the bounties on the thylacine progressively yielding fewer skins (attesting to a declining population), Paddle finally notes the efficacy of the Tasmanian political lobby against attempts to grant the species protected status. Contrasted with the successful assessment and protection of threatened koala populations in Queensland, the failure to protect thylacines until July 1936 came too late for the last known thylacine in captivity (with no further sightings in the wild officially confirmed since). This last individual at Hobart Zoo herself suffered from neglect (though his or her sex is itself a source of speculation, and the lack of records about the identity of this individual speaks further of this neglect), leading to her death from exposure on 7th September 1936 (Paddle 2002, p. 195).

The literature on the thylacine demonstrates the wider utility of analogues for reconstructing (a) the lives and habits of extinct prehistoric creatures; and (b) poorly understood extant and historically extinct species. Whilst the best source of information for the reconstruction of the life of this species is the set of anecdotal reports from colonial settlers, more productive analogies, extricated from these all too often scapegoating accounts, can be drawn with extant species (Jones and Stoddart 1998, p. 239). Some analysis that seems to confirm Paddle's scepticism towards the portrayal of the thylacine as a threat to livestock (particularly sheep) is based on the examination of museum specimens of the animal. Based on analysis of tooth shape and other physical features, Jones and Stoddart (1998), for example, suggest ready similarities between the thylacine and quolls and 'some smaller canids such as foxes and hyaenids' (Jones and Stoddart 1998, p. 241). Furthermore, they note that the thylacine's muzzle was relatively elongated in comparison with other marsupials, reinforcing the likeness with foxes rather than with wolves. This distinction is meaningful insofar as the length of an animal's muzzle indicates its likely bite force. The longer the snout, the weaker the bite force. Thus, the physiological similarity between thylacine and fox (as opposed to tiger or wolf) suggests a similarity of hunting habits and prey: in this case, a solitary 'pounce-pursuit' predator of small prey (birds and small mammals) (Jones and Stoddart 1998, pp. 243–344).

If the thylacine was not inferior, but strongly analogous, to extant placental mammals, its distinction from the latter is at least explicable in terms divested of the essentially colonialist outlook of placental chauvinism. Janis and Figueirido (2014) point to limits on marsupial carnivore specialization due to the reproductive cycle of marsupial mammals: 'One suggestion has been that marsupials cannot become as specialized in their forelimb anatomy as placentals, due to the constraints of parturition, when they have to make the climb to the pouch' (Janis and Figueirido 2014, p. 1337). They note further that:

> it appears that Plio-Pleistocene to Recent thylacines, at least, were the only medium sized carnivores in the faunas (with the exception of the highly specialized marsupial lions), and so were never under any selective pressure to specialize their type of predatory behavior, although their craniodental morphology was clearly specialized for meat-eating. (Janis and Figueirido 2014, p. 1337)

Meanwhile, Prowse et al. (2013), utilizing a model of 'population viability analysis' confront the contention that disease rather than the impact of European settlement in Tasmania precipitated the thylacine's decline. Normally used to assess risks towards still-extant populations, they apply their model to the historical case of the thylacine to shed light, retrospectively, on the possible precipitants of its extinction. They conclude that there were multiple drivers of this species' extinction, but that the negative effects of European settlement were so significant that, even without some sort of disease affecting the population that might be attributable to non-human causes, 'the species was committed to extinction' (Prowse et al. 2013, p. 363).

Thus, the image of the thylacine as threat also demonstrates the overriding shortcomings of the use of analogues to define this species. In response to some of Paddle's analysis of the historical record of the thylacine, Carol Freeman suggests a further critical response to parts of the visual record in particular. For example, Freeman examines in close detail a set of photographs by the amateur naturalist, Harry Burrell, held at the Australian Museum. The photographs, depicting a thylacine with what appears to be a freshly caught chicken in its mouth, were distributed at the time (in the 1920s) to reflect, and indeed reinforce, the idea that thylacines were a threat to livestock (see Paddle 2002 and Freeman 2005 for the images themselves). Whilst Paddle cites one of the photographs for its distribution in edited form that obscures what is clearly wire fencing in the background, showing that it is certainly not an image of a wild animal, Freeman goes one step further and suggests that the photographs in fact depict two dead animals: a taxidermy specimen of a thylacine posed with a dead chicken in its mouth. As Freeman concludes:

> the photograph of the thylacine with a chicken indeed shows a specimen rather than a live animal, on the grounds that the image is cropped to omit any indications of an enclosure; the larger scene shows it has been constructed to mimic a 'bush' setting; the figure is improbably still and in focus and has unrealistically rigid, angular legs and feet; there are signs of retouching on the tail and body that imply imperfections in the skin have been disguised; and the torso shows signs of wasting and dull fur associated with taxidermy. (Freeman 2005, p. 8, online source)

The above representations of the thylacine in the scientific and historical literature are effectively a series of analogues of a now lost species. From the stereotype of a silent, primitive hunter threatening the livestock of colonial settlers, the retrograde example of a doomed evolutionary predecessor to modern mammals, or the bad analogue of northern hemisphere carnivores, to a final caricature in staged photographs of a mounted specimen, the thylacine was perhaps never really *seen*.

2 Reverberations

In *De l'Habitude* (1838), Félix Ravaisson recommends that the study of life commences with a study of processes of the body that we can witness for ourselves, rather than through the analysis of the external form or 'geometry' of the body. Examples of such internal alterations in the body can be found in the acquisition of physical habits, indicating the oscillation between levels of receptivity of the body (to external stimuli) and the emergence of spontaneous movements or actions regardless of the presence of external stimuli. Open up a vein at regular intervals over time; eventually, one's blood will converge at the wound spontaneously. The initial trauma, the wound, persists as inflammation or hypersensitivity. Such phenomena of the organic body recur without the intervention of any determinate cause. Reaction transforms into spontaneous action (Ravaisson-Mollien 1984, p. 17). Compare this description of organic tendency with the phenomenon of stereotypy, defined as 'involuntary, coordinated, patterned, repetitive, rhythmic, and purposeless but seemingly purposeful movement'[3] (Muthugovindan and Singer 2009, p. 131).

Then consider some examples of British artist Louise Pallister's engagement with the problem of 'zoochosis,' or stereotypy identified in confined animals, typically in zoos, as a real aspect of animal life. Pallister's works help to conceptualize lived time in some of its more negative and challenging manifestations in our global context of declining habitats for non-human species, and concurrent declines and extinctions of some of those species. What tools might we employ 'to think' (or 'to draw') the dynamics of declining life forms? If we think about the experiences of these groups of animals in the light of a philosophy of process or becoming, these two modes or aspects of time, stereotypy or 'zoochosis' and extinction, pose a challenge to the accuracy and authority of our forms of representation. In the first case, animals that are typically accustomed to large territorial ranges (that happen to be the large 'charismatic' fauna favoured for zoological 'collections') are prone to develop stereotypic movements as outward symptoms of psychological damage. A number of Pallister's works resist the romanticized depictions of wild animals often found in 'wildlife art.' In 'Stereotypy,' Pallister uses the basic photographic process of cyanotype printing (an innovation of Anna Atkins in her use of the process in 1843 to catalogue plant specimens) to depict the repetitive motions of a caged tiger. The pacing tiger is both a

[3]And further: 'Other supplementing criteria have included nonreflexive rhythmic movements, which last for seconds to minutes, occur in clusters, and appear many times per day; associated with periods of stress, excitement, fatigue, or boredom; briefly affected by an ongoing excited activity but do not interfere with the ability to perform tasks; and readily suppressible by distraction or initiation of another action. On the basis of DSM-IV criteria, the repetitive, nonfunctional behaviors must be present for at least 4 weeks and markedly interfere with normal activities or possibly cause self-injury' (Muthugovindan and Singer 2009, p. 131).

beautiful and a sick animal. Pallister poses the question of the representation of other species within the context of increased stresses on the environment as a question of disconnection, asking 'how can the positive act of mark-making describe negative states of being? How best to describe an extinction or depict a creature not present or no longer intrinsic to our experience? Certainly not with detailed observational drawings' (Pallister 2016). In her sketches of the extinct thylacine, such as 'Extinction' (2014) and 'Sisyphus' (2014), Pallister places greater emphasis on erasure than on singular or analogical anatomical features of the lost species:

> My work focuses on the negative spaces of vanished and vanishing animals. The sketchy silhouettes in my drawings indicate where extinct or endangered animals and birds should be; manifestations of their absence. Elsewhere, erasing sections of drawing eventually led me to make a stop-motion animation. I drew frames from the film of the caged thylacine, photographed the successive drawings, then erased and redrew, in a repeated passage from paper to camera, reminiscent of the captive animal itself. Three hundred erased drawings later I had my animation: *'Caged'* (2014), a large-scale charcoal thylacine eternally circling a cage, a Sisyphean prisoner. In this way I sought to span the space between the original film and our experience of it: the creature comes and goes, fluctuating between a world that held such an animal and one that doesn't.' (Pallister 2016)

I would suggest some continuity here with William James' proposal for a truly radical empiricism that requires that we exclude from its statements anything that is not directly experienced, whilst also taking care *not* to leave out any element of things that *is* directly experienced (James 1996, p. 42). The experience of the extinction of particular species might be articulated in this way; not simply in terms of the removal of a species from the biota, but the experience of its absence or death.[4] For James, what we call 'consciousness' is not some mysterious entity or special mode of being. It is rather to be defined in terms of our direct experiences and the relations between this multitude of experiences

[4] See Thom van Dooren (2014) and Rose et al. (2017).

(James 1996, p. 25). Anticipating Whitehead's focus on prehensions as the orientation of minimal experiences towards one another, James tries to alter the discourse of our philosophical enquiry to focus less on abstract statements such as 'I think' to more concrete expressions such as 'I breathe' (James 1996, p. 37). There is only pure experience, where thing and perception coincide, in which both partake of the same multiplicities of qualities (hotness, blueness, and so on). Noting objections to this focus on affections as a more 'spiritual' interpretation of the physical elements that make up 'pure experience,' James responds:

> Take a mass of carrion, for example, and the 'disgustingness' which for us is part of the experience. The sun caresses it, and the zephyr wooes it as if it were a bed of roses. So the disgustingness fails to *operate* within the realm of suns and breezes,—it does not function as a physical quality. But the carrion 'turns our stomach' by what seems a direct operation—it *does* function physically, therefore, in that limited part of physics. [...] Our body itself is the palmary instance of the ambiguous. Sometimes I treat my body purely as a part of outer nature. Sometimes, again, I think of it as 'mine.' I sort it with the 'me,' and then certain local changes and determinations in it pass for spiritual happenings. Its breathing is my 'thinking,' its sensorial adjustments are my 'attention,' its kinesthetic alterations are my 'efforts,' its visceral perturbations are my 'emotions.' (James 1996, p. 153)

In *A Pluralistic Universe*, William James surveys the current state of academic philosophy (as it was at the turn of the twentieth century) and cautions against the loss of connection between the discipline and 'the open air of human nature' (James 1997, p. 17). Should we seek to restore that lost connection, he continues, we would find that 'Nature, more demonic than divine, is above all things *multifarious.* So many creatures that feed or threaten, that help or crush, so many beings to hate or love, to understand or start at—which is on top and which subordinate? Who can tell?' (James 1997, p. 22). The division in philosophy that James explores, contrasted with the reductionist trajectory of materialism, is between the possibilities of 'spiritualist philosophy' along the lines of dualistic theism, on the one hand, and pantheism on the other. Divided again, a pantheist universe might take either a monistic form or a pluralistic one. James' radical empiricism, following the pluralistic line, 'allows that the absolute sum-total of things may never be actually experienced or realized in that shape at all, and that a disseminated, distributed, or incompletely unified appearance is the only form that reality may yet have achieved' (James 1997, p. 44). We thus arrive at a view of a pluralistic universe that is 'a turbid, muddled, gothic sort of an affair, without a sweeping outline and with little pictorial nobility' (James 1997, p. 45).

Two thinkers whose work figures most significantly within James' lectures are Gustav Fechner and Henri Bergson. From Fechner, James derives a methodological approach to the complexity of real experience and its relations to the world at large. He cites Fechner's peculiar facility with analogical reasoning, and an appreciation of its limits, as the means to gain new insights into longstanding problems in the history of philosophy. For example, his discussion of the nature of God's mind forces insight into the illegitimate application of anthropomorphic qualities onto non-anthropomorphic things. We might, as James suggests, infer from the connection of our minds to our bodies that God's mind is embodied (James 1997, pp. 151–152). However, the limits of the analogy are clear if we take care to remember the role of evolutionary adaptation in shaping our bodies. Other beings may well be embodied minds, but their exposure to their specific environments will undoubtedly shape for them bodies that are quite different to ours. We can thus take the hypothetical case of a divine body to help us think otherness, and then apply that reasoning to a consideration of what counts as a well-adapted or 'advanced' form of embodiment.

James' admiration for Fechner's use of analogy to explain his otherwise abstract ideas is understandable when one reads *The Little Book of Life After Death* (first published in 1834).[5] Whilst my own focus

[5]Bergson defends James' use of the 'curious theory' of Fechner: 'The powerful feelings which stir the soul at certain special moments are forces as real as those that interest the physicist; man does not create them any more than he creates heat or light' (Bergson 1992, p. 212). And further, concerning the misinterpretation of James' book *The Varieties of Religious Experience* as simply a psychology of religious feeling: 'The truth is that James leaned out upon the mystic soul as, on a spring day, we lean out to feel the caress of the breeze on the cheek, or as, at the sea-side, we watch the coming and going of sail-boats to know how the wind blows. Souls filled with religious enthusiasm are truly uplifted and carried away: why could they not

on the relation between process philosophy and the sciences precludes discussion of theological matters, Fechner's speculative diversion into the possibility of life after death (one that is inspired, if not wholly in tune with, Christian doctrine) is eloquent and compelling. The tale Fechner tells of the three lives of a human being, the first stage commencing in the womb, the second in life, and the third in life after death is both moralistic in its teaching and indicative of a panpsychist schema extending beyond the moment of biological death. 'Whatever any one has contributed during his life, of creation, formation, or preservation, to the sum of human idealism,' he writes, will persist in the thoughts and actions of others, thereby constituting at least one dimension of survival after death (Fechner 1905, p. 10). The mode of this survival is one of degrees of influence, spanning one's immediate actions to the furthest effects of a conversation or a piece of writing by the most unassuming individual, but more powerfully by a great thinker ('a Goethe, a Schiller, a Napoleon, a Luther') (Fechner 1905, p. 13). Such influence can be for the good of others, driving creative, collective action, or it can be a force of persisting conflict that manifests itself 'in the rivalry of systems, in sectarian hatred, in wars and revolutions between princes and people, and the nations among each other' (Fechner 1905, p. 36). The mechanism of influence is relayed in Fechner's characteristic analogical manner:

> Although the undulating circle which a sinking stone leaves behind it in the water creates, by its contact, a new circle around every rock which still projects above the surface, it still retains in itself a connected circumference which stirs and carries all within its reach; but the rocks are only aware of the breaking of the perfect line. We are just such ignorant objects, only that we, unlike fixed rocks, while even still in life, shed about us a continuous flow of influence which extends itself not only around others but within them. (Fechner 1905, pp. 11–12)

---

enable us to experience directly, as in a scientific experiment, this uplifting and exalting force?' (Bergson 1992, pp. 212–213).

The moral then begins to give way to the metaphysical when Fechner's image of the place of plants on the Earth reflects the nature of the relation between parts and whole. Each individual is defined and confined by what it feels is its own purpose: for example, a plant whose purpose is 'to drink in light and air, to prepare fragance and color for its own adornment, to play with beetles and bees' (Fechner 1905, p. 22). However, each individual plant contributes to processes that help the earth's ecosystems to 'exhale, breathe, [and] weave for itself a green garment' on which all other living things depend for their subsistence (Fechner 1905, p. 22). This cascade of interrelations extends to all living things, and it includes as part of it the complex of human consciousnesses and the transmission of ideas between individuals and societies across generations. Fechner's simple diagram of intersecting circles (Fechner 1905, p. 24) seeks to communicate something of this interconnectedness, whilst the ensuing pages of the text, in particular a number of footnotes referring to some of Fechner's formal writings, begin to draw out the more microscopic processes of human thought. Explaining the dynamic of survival of a thought in action or the communication of an idea, Fechner writes in a footnote:

> Whether one attributes nervous energy to a chemical or an electrical process, one must still regard it, if not simply as the play of the vibration of minutest atoms, yet as in the main excited or accompanied by this, whereby the imponderable has a larger part than the ponderable. Vibrations, however, can only apparently expire by extending themselves into their environment, or if indeed they disappear for a time through translation of their living strength into so-called elasticity, yet, according to the law of the conservation of energy, they await a revival in some other form. (Fechner 1905, pp. 58–59)

The strength of this transmission of energy is tempered by 'the law of the threshold of consciousness' that regulates both the capacity of an individual body to exert its influence on itself and its surroundings; if a body is weakened to a certain point (from illness, or simply in sleep), it can pass below its threshold of consciousness. The more the individual (body) extends its actions more widely, the closer it falls to its threshold, or the limits of its possible action. In a further footnote, Fechner expands this threshold concept to explain the dynamic of influence between different individuals, and indeed the whole of life on earth. A single consciousness can survive its projection into the world at large (in Fechner's speculations on the transition from mortal life into the after-life), analogously to its capacity to survive according to the limits (the threshold) of its own body.[6]

In his introduction to this fourth edition of Fechner's book (published in 1905), William James summarizes the image of consciousness that Fechner's writing conveys. Each conscious being is so immersed in its own action and the sense of its own purpose that it is barely able to discern that it and the universe are 'coeternal aspects of one self-same reality, much as concave and convex are aspects of one curve' (James, in Fechner 1905, p. xiii). In this expanded schema, 'our bodies are just wavelets on the surface of the earth. We grow upon the earth as leaves grow upon a tree, and our consciousness arises out of the whole earth consciousness, which it forgets to thank, just as within our consciousness an emphatic experience arises, and makes us forget the whole background of experience without which it could not have come' (James, in Fechner 1905, p. xv).

If we return to William James' reading in *A Pluralistic Universe*, expanding our view further outwards still, James cites Fechner's expansive analogical description of the Earth as a gigantic organism. As we saw earlier, in the combined insights of Lovelock and Margulis (together with Bateson's discussions of the 'organism-plus-environment'), references to an organismic model of the Earth's ecosystem suggest alternative ways to think about complex forms of interdependency in the realms of the

[6]See, in particular, Fechner's detailed footnotes in which he continues his explanation of the dynamics of thresholds: 'If the psychophysical life-wave (to continue the use of this concise expression) of man, made up of components of the most manifold sort, should spread out into a world which contained only different components, then, indeed, must it be assumed that it, in its extension, would fall below the combined-threshold here under consideration. Since, however, the psychophysical undulatory sea of the universe, among its other components, comprehends also such as are like to those of the human life-wave, and indeed of the most varying height or intensity, therefore such as already rise above or come near the level of the combined-threshold and are only raised still higher by the similar ones which join them, so is the result of the above speculation placed on a somewhat more solid basis' (Fechner 1905, pp. 82–83).

chemical and biological sciences. From the perspective of process philosophy, Fechner's analogy allows us to think the interconnection of complex forms: 'As the total bearing of any animal is sedate and tranquil compared with the agitation of its blood corpuscles, so is the earth a sedate and tranquil being compared with the animals whom she supports' (James 1997, p. 157). Fechner's image thus inspires a pluralistic panpsychism in which the significant relation described is that between different forms of consciousness, from the more complex or 'inclusive' to the more limited (James 1997, p. 168). Encompassed by this 'earth-soul' all forms of consciousness are interwoven, but the degree to which they each 'know' or 'take account of' one another is different in each case. Two key implications follow. Firstly, the Fechnerian earth-soul suggests how subtraction, or more personally, *loss* might be experienced on a global level:

> Fechner likens our individual persons on the earth unto so many sense-organs of the earth's soul. We add to its perceptive life so long as our own life lasts. It absorbs our perceptions, just as they occur, into its larger sphere of knowledge, and combines them with the other data there. When one of us dies, it is as if an eye of the world were closed, for all *perceptive* contributions from that particular quarter cease. (James 1997, p. 171)

Secondly, taking up James' reading of the philosophy of Henri Bergson, we can develop our account of the nature of interconnection that subsists between the parts of this global system. From Bergson's philosophy, James takes up the problem of the constitution of time and its harmony (or otherwise) with conscious experience. The suggestion of real time as somehow an abstract container, in the manner of abstract space, leads, as Bergson illustrates, to the paradoxes of Zeno. If, instead, we assume that real time approximates to our real experiences and is something that grows 'by units of duration of determinate amount, just as our perceptions of it grow by pulses,' then the paradoxes would disappear (James 1997, p. 231). Conscious experience affords us immediate access to the constitution of time, in which 'all our sensible experiences, as we get them immediately, do thus change by discrete pulses of perception, each of which keeps us saying "more, more, more," or "less, less, less," as the definite increments or diminutions make themselves felt' (James 1997, p. 231). Time is not simple continuity, or flow, but rather drops or pulses of becoming. Invoking Fechner's concept of thresholds, which describe the discrete changes in our perceptual experience, James follows the line of Bergson's thought in tracing the movement from the confusion of past and present in experience to the artificial precision of concepts applied to it:

> All *felt* times coexist and overlap or compenetrate each other thus vaguely, but the artifice of plotting them on a common scale helps us to reduce their aboriginal confusion, and it helps us still more to plot, against the same scale, the successive possible steps into which nature's various changes may be resolved, either sensibly or conceivably. (James 1997, p. 232)

We can, James continues, compare the intellectual act of analysing motion into discrete, successive spaces with the image of a man with vertigo, who can gradually orient and steady himself in relation to his sensations in order, eventually, to walk without staggering (James 1997, p. 233). The confused multiplicity of sensory experience is continually subject to the organizing work of our memories and expectations that are directed towards harnessing our external environment for potential action.

James' criticisms of abstract conceptualizations of the flow of reality are reinforced in Bergson's identification of intellectual work with useful action. Concepts, doing practical, rather than speculative work, necessarily fall short of reality as becoming. James' assessment of the shortcomings of the concept hinges on the dissonance between an intellectualist logic of exclusion (a thing is what it is, by virtue of excluding what it is not) and the flow of reality which seems to be 'its own other' (James 1997, p. 109). Bergson's contention concerning the practical utility of concepts frees us from the need to think reality, or real time, in line with such abstract frameworks. Rather, a return to the examination of real experience might yield insights into the ways in which the 'concrete pulses' of reality are 'their own others' (James 1997, p. 109). As James explains further, it is not simply that we are out of step with reality, but rather that our concepts, whilst employed for the purposes of practical action, in a complementary way do in fact also afford us the capacity to *engage with* our surroundings, to 'get into' things, which would, conversely, imply that we are able 'to get out of one's self and be one's other, which is self-contradictory, etc. Meanwhile each of us actually *is* his own other to that extent, livingly knowing how to perform the trick which logic tells us can't be done' (James 1997, p. 259). In his tribute entitled 'On the Pragmatism of William James,' Bergson explains further this discord with reality through an image of the relation between conceptual and real knowledge as akin to the relation between actors on a stage and our experiences of our everyday lives. The words and actions on the stage are economical and contrived to converge on a resolution, be it a happy or a tragic one, in the final act of the play. In our real lives, the confusion and excessive superfluity of everyday conversations and actions have no apparent order that might be unified into a greater purpose. Reality is 'redundant and superabundant' (Bergson 1992, p. 210).

In reading the work of other thinkers with a 'post-mortem method' of reconstructing a philosophy from its rarefied elements, one is in danger of merely re-enacting the mimicry of a life on the stage, and stumbling through one's reading 'like a myopic ant over a building, tumbling into every microscopic crack or fissure, finding nothing but inconsistencies, and never suspecting that a centre exists' (James 1997, p. 263). James advocates for a pluralist view of reality, and resists being drawn into the fight to wield the word 'rationality,' lest one be accused of 'irrationality' (as indeed Bergson has been). James proposes the opposition between 'foreignness' and 'intimacy,' where 'the notion of the "one" breeds foreignness and that of the "many" intimacy' (James 1997, p. 321). The pluralist doctrine advocates a view of reality in which its parts are externally related, rather than existing in all-encompassing or dominating relations to one another: 'Things are "with" one another in many ways, but nothing includes everything, or dominates over everything. The word "and" trails along after every sentence. Something always escapes' (James 1997, p. 321). A monistic view of reality, on the other hand, would necessarily require some concept of an ultimate 'enveloper' (as James observes of the terminus of Fechner's view of the earth-soul), but pluralism states that:

> nothing real is absolutely simple, that every smallest bit of experience is a *multum in parvo* plurally related, that each relation is one aspect, character, or function, way of its being taken, or way of its taking something else; and that a bit of reality when actively engaged in one of these relations is not *by that very fact* engaged in all the other relations simultaneously. (James 1997, pp. 322–323)

In this opposition between the 'all-form' of monism and the 'each-form' of pluralism, the latter understood as a 'multiverse' conveys an image of the universe as a lattice of interconnecting parts each existing 'in some possible or mediated connexion, with every other part however remote, through the fact that each part hangs together with its very next neighbours in inextricable interfusion' (James 1997, p. 325). In the light of James' use of Fechner's world-soul, configured as a 'multiverse,' to think *ecologically* about the relations between different parts of the material universe, the model of organic regulation, scaled up to a global level allows us to think about the functions and reactions of Gaia on a psychological level. The effects of species loss may well be redressed by the Gaian system, but, in the spirit of a radical empiricism, the description of this dynamic of compensation must also incorporate the experience of loss. The loss of a species reverberates across different levels of the system. Traumas or wounds have a tendency to return and repeat. Concern over the loss of species, and the repetitions and revisions of our analysis of the lives of the absent, are the reverberations of the trauma of those losses and fosters in us the desire to retreat from the world when it is most in need of our attention.

References

Attard, M.R.G., U. Chamoli, T.L. Ferrara, T.L. Rogers, and S. Wroe. 2011. Skull Mechanics and Implications for Feeding Behaviour in a Large Marsupial Carnivore Guild: The Thylacine, Tasmanian Devil and Spotted-Tailed Quoll. *Journal of Zoology* 285 (4): 292–300. https://doi.org/10.1111/j.1469-7998.2011.00844.x.

Bergson, Henri. 1992. *The Creative Mind: An Introduction to Metaphysics*, trans. Mabelle L. Andison. New York: Citadel Press.

Darwin, Charles. 1996. *The Origin of Species*. Oxford: Oxford University Press.

Fechner, Gustav. 1905. *The Little Book of Life After Death*, trans. Mary C. Wadsworth. Boston: Little, Brown.

Freeman, Carol. 2005. Is This Picture Worth a Thousand Words? An Analysis of Harry Burrell's Photograph of a Thylacine with a Chicken. *Australian Zoologist* 33 (1): 1–15. https://doi.org/10.7882/AZ.2005.001.

Gould, Stephen Jay. 1991. *Bully for Brontosaurus*. London: Hutchinson Radius.

James, William. 1996. *Essays in Radical Empiricism*. Lincoln and London: University of Nebraska Press.

James, William. 1997. *A Pluralistic Universe*. Lincoln and London: University of Nebraska Press.

Janis, Christine M., and Borja Figueirido. 2014. Forelimb Anatomy and the Discrimination of the Predatory Behavior of Carnivorous Mammals: The Thylacine as a Case Study. *Journal of Morphology* 275 (12): 1321–1338. https://doi.org/10.1002/jmor.20303.

Jones, Menna E., and D. M. Stoddart. 1998. Reconstruction of the Predatory Behaviour of the Extinct Marsupial Thylacine (*Thylacinus Cynocephalus*). *Journal of Zoology* 246 (2): 239–246. https://doi.org/10.1111/j.1469-7998.1998.tb00152.x.

Muthugovindan, Deivasumathy, and Harvey Singer. 2009. Motor Stereotypy Disorders. *Current Opinion in Neurology* 22 (2): 131–136. https://doi.org/10.1097/WCO.0b013e328326f6c8.

Paddle, Robert. 2002. *The Last Tasmanian Tiger: The History and Extinction of the Thylacine*. Cambridge: Cambridge University Press.

Pallister, Louise. 2016. Extinction and the Image. The Learned Pig. http://www.thelearnedpig.org/extinction-and-the-image/3908. Accessed 12 September 2019.

Prowse, Thomas A.A., Christopher N. Johnson, Robert C. Lacy, Corey J.A. Bradshaw, John P. Pollak, Michael J. Watts, and Barry W. Brook. 2013. No Need for Disease: Testing Extinction Hypotheses for the Thylacine Using Multi-Species Metamodels. *Journal of Animal Ecology* 82 (2): 355–364. https://doi.org/10.1111/1365-2656.12029.

Ravaisson-Mollien, Félix Lacher. 1984. *De l'Habitude; La Philosophie en France au XIX Siècle*. Paris: Librairie Arthème Fayard.

Rose, Deborah Bird, Thom van Dooren, and Matthew Chrulew. 2017. *Extinction Studies: Stories of Time, Death, and Generations.* New York: Columbia University Press.
Van Dooren, Thom. 2014. *Flighthways: Life and Loss at the Edge of Extinction.* New York: Columbia University Press.
Wemmer, Christen M. 2002. Opportunities Lost: Zoos and the Marsupial That Tried to Be a Wolf. *Zoo Biology* 21: 1–4.

Index

W. Khandker, *Process Metaphysics and Mutative Life*,
Palgrave Perspectives on Process Philosophy,
https://doi.org/10.1007/978-3-030-43048-1

The manufacturer's authorised representative in the EU is Springer Nature Customer Service Centre GmbH, Europaplatz 3, 69115 Heidelberg, Germany. If you have any concerns regarding our products, please contact ProductSafety@springernature.com

Printed and bound by CPI Group (UK) Ltd, Croydon, CR0 4YY
30/09/2026
02225461-0002